T0200065

HAS SCIENCE
FOUND GOD?

VICTOR J. STENGER

HAS SCIENCE FOUND GOD?

The Latest Results in the Search for
Purpose in the Universe

 Prometheus Books

59 John Glenn Drive
Amherst, New York 14228-2197

Published 2003 by Prometheus Books

Inquiries should be addressed to
Prometheus Books
59 John Glenn Drive
Amherst, New York 14228–2197
VOICE: 716–691–0133, ext. 207
FAX: 716–564–2711
WWW.PROMETHEUSBOOKS.COM

07 06 05 04 03   5 4 3 2 1

Library of Congress Cataloging-in-Publication Data

Stenger, Victor J., 1935–
    Has science found God? : the latest results in the search for purpose in the
universe / Victor J. Stenger.
        p. cm.
    Includes bibliographical references and index.
    ISBN 1–59102–018–2
    1. Religion and science. 2. Atheism. I. Title.

BL240.3 .S74 2002
291.1'75—dc21

2002036717

To my granddaughters Katelyn Sachiko Stenger,
Lucy Anna Green, and Zoe Aiko Stenger,
children of the new millennium

CONTENTS

Preface   9

1. The Bells of St. Mary's   29

2. The Menace of Darwinism   43

3. No Reason to Believe   77

4. Messages from Heaven   99

5. The God of Falling Bodies   131

6. The Uncreated Universe   141

7. The Other Side of Time   163

8. The Laws of the Void   187

9. Absence of Evidence   219

10. The Breath of God   261

11. The Premise Keepers   307

12. The Godless Universe   339

Appendix A. The Planck Length   351

Appendix B. The Lifetimes of Stars   354

Appendix C. The Entropy of the Expanding Universe   356

Index   359

PREFACE

Scientific criticism has no nobler task than to shatter false beliefs.
—Ludwig von Mises, 1932

WHERE WAS GOD ON SEPTEMBER 11?

The original manuscript of this book was mailed to the publisher on September 11, 2001, the day three thousand lives ended abruptly at the hands of Muslim terrorists. While many have assumed that their motives were political, there can be no denying that their fanaticism was fueled by their faith. They fully believed they were acting in the name of God.

Political figures, including the president of the United States, and many religious leaders went to great lengths to play down the role of religion in the tragedy. Generally the response was that this was an aberration, that Islam is a "religion of peace." Much public prayer followed and preachers sought to comfort their flocks by assuring them that God still loved them. Some explained that it was the devil, not God, who was responsible for the horror. But others suggested that God was sending America a message.

On Pat Robertson's television program, the *700 Club*, on

September 13, 2002, the Rev. Jerry Falwell asserted: "What we saw on Tuesday, as terrible as it is, could be miniscule if, in fact, God continues to lift the curtain and allow the enemies of America to give us probably what we deserve." Robertson replied, "Well, Jerry, that's my feeling. I think we've just seen the antechamber to terror, we haven't begun to see what they can do to the major population." Falwell then added:

> The ACLU has got to take a lot of blame for this. And I know I'll hear from them for this, but throwing God . . . successfully with the help of the federal court system . . . throwing God out of the public square, out of the schools, the abortionists have got to bear some burden for this because God will not be mocked and when we destroy 40 million little innocent babies, we make God mad. . . . I really believe that the pagans and the abortionists and the feminists and the gays and the lesbians who are actively trying to make that an alternative lifestyle, the ACLU, People for the American Way, all of them who try to secularize America . . . I point the thing in their face [sic] and say you helped them.[1]

In the resulting public uproar, Falwell apologized for his comments and Robertson distanced himself from them. However, both continue to blame secularism for evils in American society and, despite his disclaimer, Falwell's original statement represents a widely held view often expressed among conservative Christians.

Falwell was echoing a similar statement made by Anne Graham Lotz, the daughter of evangelist Billy Graham, that same morning on the CBS *Early Show*:

> I would say also for several years now Americans in a sense have shaken their fist [sic] at God and said, God, we want you out of our schools, our government, our business, we want you out of our marketplace. And God, who is a gentleman, has just quietly backed out of our national and political life, our public life. Removing his hand of blessing and protection.[2]

Not everyone in the religious community viewed secularism as at fault. Several clergymen who were on the actual scene at the

World Trade Center found themselves unable to hold religion blameless. In a Public Broadcasting System Frontline television documentary shown as the first anniversary approached,[3] Orthodox Rabbi Brad Hirschfield ruefully admitted that "religion drove those planes into those buildings." Monsignor Lorenzo Albacete, Catholic priest and theology professor, eloquently described his feelings when we first heard the news:

> From the first moment I looked into that horror on September 11, into that fireball, into that explosion of horror, I knew it. I knew it before anything was said about those who did it or why. I recognized an old companion. I recognized religion. Look, I am a priest for over thirty years. Religion is my life, it's my vocation, it's my existence. I'd give my life for it; I hope to have the courage. Therefore, I know it.
>
> And I know, and recognized that day, that the same force, energy, sense, instinct, whatever, passion—because religion can be a passion—the same passion that motivates religious people to do great things is the same one that that day brought all that destruction. When they said that the people who did it did it in the name of God, I wasn't the slightest bit surprised. It only confirmed what I knew. I recognized it.
>
> I recognized this thirst, this demand for the absolute. Because if you don't hang on to the unchanging, to the absolute, to that which cannot disappear, you might disappear. I recognized that this thirst for the neverending, the permanent, the wonders of all things, this intolerance or fear of diversity, that which is different—these are characteristics of religion. But I knew that there was no greater and more destructive force on the surface of this Earth than the religious passion.
>
> At least some form of religion did, and pretending that it wasn't is a very dangerous lie we tell ourselves.

A young Episcopal priest, Joseph Griesediek, found his faith shaken as he walked amid the debris:

> Prior to September 11, the face of God for me was one that was strong, secure, consistent; a face that, while at times seemed distant, can more or less be counted on to be there, who kept things

in order—the sun would come up, the sun would go down—
who would provide, could be counted on.

After September 11, the face of God was a blank slate for me.
God couldn't be counted on in the way that I thought God could
be counted on. That's what I felt as I stood on Ground Zero. God
seemed absent. It was frightening, because the attributes that I
had depended upon in the past, when thinking about the face of
God, had all been stripped away. I was left with nothing but that
thing we call faith. But faith in what? I wasn't so sure.

The face of God after September 11 is much more of a mystery that it ever was, a mystery that is still unfolding. I see
glimpses of this face of God when I talk to people who have also
been at Ground Zero, and who know that there was some kind of
force keeping us together, working to insure some kind of survival. But when that face of God that I used to see [is] proclaimed
[by] people who weren't at Ground Zero, it doesn't speak to me
any longer. In some ways, I believe that on Ground Zero I grew
up, and part of that growing up is truly grasping that which can't
be grasped—the mystery of God—a face that often eludes us,
and frustrates me.

The face of God is no mystery. It is as blank as Rev. Griesediek
sees it to be. The face of God was not seen on September 11, 2001,
nor has it been seen in the many other great catastrophes that have
struck humanity throughout history. The face of God was invisible
when 2,000 people perished in an earthquake in India earlier that
year. It was not seen by the half-million Americans who died of the
flu in 1918, nor by the many millions who were killed in the twentieth century's great wars. The face of God was just as blank at
Auschwitz as it was at Ground Zero in Manhattan. It is not seen by
the child dying of leukemia.

Despite the apparent absence of a loving creator in so many
tragedies, billions of people cling to faiths in which a personal God
sustains the universe. Most of these are the poor and uneducated
of the undeveloped world, easily carried away by the magic and
ritual that religion supplies. In developed countries, religion is
gradually dying out. The one giant exception is the United States,
where the great majority of a highly educated populace still pro-

fesses belief in a supreme being. Among these, many if not most will tell you that their belief is not based on some emotional need but on their own observations of the world around them. They insist that these observations show evidence for a God that very much resembles that found in the traditions of the Abrahamic religions—Judaism, Christianity, and Islam. They find support for their beliefs in an immense literature that has been produced in recent years claiming that the face of God can be seen—in science.

SCIENCE AND THE SPIRITUAL QUEST

In huge letters, the July 20, 1998, cover of *Newsweek* announced to the world, "Science Finds God."[4] Inside, the cover story reported on a conference, "Science and the Spiritual Quest," held that summer at the Center for Theology and the Natural Sciences in Berkeley, California. The scientists and theologians who gathered at the meeting were virtually unanimous in agreeing that science and religion were now converging, and what they were converging on was God. South African cosmologist and Quaker George Ellis expressed the consensus, "There is a huge amount of data supporting the existence of God. The question is how to evaluate it."

The *Newsweek* story noted, "The achievements of modern science seem to contradict religion and undermine faith." However, "For a growing number of scientists, the same discoveries offer support for spirituality and hints at the very nature of God." According to the author, science writer Sharon Begley, "Physicists have stumbled on signs that the cosmos is custom-made for life and consciousness." Big bang cosmology, quantum mechanics, and chaos theory all are interpreted as "opening a door for God to act on the world."

Distinguished astronomer Alan Sandage told the Berkeley gathering that he was once a nonbeliever, but has been driven by his science to the conclusion that "the world is much more complicated than can be explained by science." He added, "It is only through the supernatural that I can understand the mystery of existence."

In his story on the Berkeley meeting, George Johnson of the

New York Times reported that Sandage complained that it was a packed crowd: "Many of the speakers have been preaching to the choir. There are no atheists on the program, only strict believers."[5] In a commentary appearing a few weeks later, Johnson noted that "religious believers seem more eager than ever to step over the line, trying to interpret scientific data to support the revealed truths of their own theology."[6]

In a recent article in the high-tech magazine *Wired*, Gregg Easterbrook asserts that modern science has a "miraculous aspect" and constitutes "supernaturalism dressed up." He accuses physicists of using "unknown physical laws" to explain the origin of the universe and theories that require "as much suspension of disbelief as any religion."[7]

The Center for Theology and the Natural Sciences is a research and teaching institute that explores "the relation between contemporary physics, cosmology, technology, environmental studies, evolutionary and molecular biology, and Christian theology and ethics."[8] It is one of a growing number of similar organizations funded either partially or fully by the John Templeton Foundation.

Established in 1987 by international investor Sir John Templeton, the foundation currently supports some 150 projects, studies, award programs, and publications worldwide. Each year, the foundation awards the Templeton Prize for Progress in Religion, which, at about one million U.S. dollars, exceeds the Nobel Prize in monetary value. Significantly, several awards have gone to physicists and other scientists for writing on how science and religion might be made compatible.

Templeton is a devout Christian, and his foundation largely reflects his philosophy with the bulk of its money going to support conferences, publications, and research by believers. The stated mission of the foundation is to "stimulate a high standard of excellence in scholarly understanding which can serve to encourage further worldwide explorations of the moral and spiritual dimensions of the Universe and of the human potential within its ultimate purpose."[9] While ultimate purpose is not questioned, Templeton scholars are free to speculate on what that may be.

Templeton grantees generally do not directly dispute scientific theories, such as evolution, or attempt to alter science's basic naturalistic methodology. Rather, they try to reconcile these with traditional religious teachings. Some Templeton scholars have managed to work unguided evolution into their theologies and look to science as a means for improving society by working hand in hand with religion.

In their willingness to operate within a framework of scientific knowledge, Templeton scholars find themselves in deep conflict with members of several other well-financed Christian groups that have also assembled in recent years to deal with issues of religion and science. These groups promote a more conservative Christian agenda with the goal of transforming both science and society so that they more closely align with their doctrinal interpretations of biblical teachings. In particular, the conservatives are aghast at the notion of unguided evolution of life on Earth. They view Darwinian evolution and the more general materialistic assumptions of science as the cause of what they perceive as the moral decay of modern society.

Leading the conservative movement in the United States are fellows of the Center for Science and Culture (CSC), an arm of the Seattle-based Discovery Institute.[10] The CSC fellows and their supporters accuse mainstream scientists of dogmatically refusing to accept the "new evidence" that signs of purposeful design in the universe can be found in scientific data from cosmology, cognitive science, and molecular biology. However, they do not press their case in scientific forums. Rather, they operate in the public and political arenas where they strive to convince laypeople and politicians that scientists need to abandon their "counterintuitive" attachment to materialism. The practical goal of the CSC is to include this "new evidence" in science curricula in the name of "fairness." This is exemplified by what they have termed the "wedge strategy." Among their stated goals are the following:

- To defeat scientific materialism and its destructive moral, cultural, and political legacies
- To replace materialistic explanations with the theistic understanding that nature and human beings are created by God

- To see intelligent design theory as the dominant perspective in science[11]

To most theistic believers, human life can have no meaning in a universe without God. With an understandable yearning for a meaning to their existence, they reject the possibility of no God. In their minds, only a purposeful universe based on God is possible and science can do nothing else but support this "truth." Continually bombarded by one side of the story, in church and in the media, most people have not given sufficient thought to the alternatives, which may not be not as bleak as they have been led to believe from the pulpit.

In *The Battle for God*, Karen Armstrong explains that historically humans evolved two ways of thinking, speaking, and acquiring knowledge, which scholars have called *mythos* and *logos*.[12] *Mythos*, or "myth," is the source of words like "mystery" and "mysticism" and derives from the Greek *musteion*: to close the eyes or the mouth. Armstrong defines it as "A mode of knowledge rooted in silence and intuitive insight which gives meaning to life but which cannot be explained in rational terms."

In the premodern world, according to Armstrong:

> Myth was regarded as primary; it was concerned with what was thought to be timeless and constant in our existence. Myth looked back to the origins of life, to the foundations of culture, and to the deepest levels of the human mind. Myth was not concerned with practical matters, but with meaning. Unless we find some significance in our lives, we mortal men and women fall very easily into despair. The *mythos* of a society provided people with a context that made sense of their day-to-day lives; it directed their attention to the eternal and the universal.[13]

Armstrong adds that myth cannot be demonstrated by rational proof and becomes a reality only when "embodied in cult, rituals, and ceremonies" which evoke within worshippers "a sense of sacred significance" that enables them to "apprehend the deeper currents of existence."

Armstrong defines *logos*, which means "word" in Greek, as "rational, logical, or scientific discourse."[14] In the modern world, mainly as a result of the great success of science, most people have come to regard *logos* as the primary method by which knowledge about reality is obtained. Thus, even those who hold personal beliefs that arose in the premodern past by means of *mythos* seek ways in which those beliefs can also be supported by *logos*. As Armstrong explains,

> In the old world, mythology and ritual had helped people to evoke a sense of sacred significance that saved them from the void, in rather the same way as did great works of art. But scientific rationalism, the source of Western power and success, had discredited myth and declared that it alone could lead to the truth.[15]

In the present book, I critically examine the attempt to apply *logos* to seek evidence for the existence of God or, more generically, for a transcendent element to the universe that has significant, observable effects. I do this from the perspective of an experimental physicist who for forty years was involved in research in elementary particle physics and astrophysics. During that period I witnessed many extraordinary discoveries in these fields that altered previous thinking. As of this writing, the theories developed to describe these discoveries provide a comprehensive picture of the nature of a purely material universe that is consistent with all existing data.

While the details of this picture will undoubtedly evolve as new data come in, and while the current theories have many theoretically unsatisfactory aspects that are bound to be improved upon, they represent the best of our current knowledge. This knowledge includes mathematical formulas and methods that allow for calculations to exquisite precision. It also includes technologies that allow these calculations to be tested empirically to equally exquisite precision. So far, the theories have passed every empirical test. No observations or measurements currently exist that can be labeled "anomalous" in the sense that they require a revision of existing theories so drastic as to render these theories inoperative at some level.

That is not to say that scientists are sitting back and letting this situation stand indefinitely. Experimentalists are working hard to find empirical anomalies that the theorists cannot explain, and theorists are working equally hard to develop new theories that will give experimentalists further frontiers to explore.

In any case, this book is neither a history of science past nor a speculation about science future. It is an application of science present to the question of the existence of a nonmaterial, divine reality that transcends the world of the senses. Such a reality has formed an important part of human belief systems for many millennia, based mainly on *mythos*. While *logos* has also been applied to the question for almost as long, this use has been largely limited to its "rational" and "logical" aspects, and not so much to its "scientific" aspect. Theology utilizes a form of *logos* that, until recently, has largely excluded science. For good reason. In the span of time since science saw its first dawn over 2,500 years ago in ancient Greece, theology has failed to empirically validate the existence of the transcendent. In fact, belief in the transcendent has been deeply undermined as scientific explanations based on a natural, material reality superseded the gods and spirits that people assumed animated the world.

Most of the scientists who ever lived are alive today, and the greatest knowledge that humanity has ever possessed is contained in the well-established observations and theories of science present. These are not speculations. These are solid facts and solid theories based on solid facts. I will show that the solid facts and theories of the present do not provide any clear signal for the presence of a reality beyond what is revealed to our senses, despite what was said at the 1998 Berkeley gathering and what is claimed on the Discovery Institute Web site.

While this book will say much about religion and theology, and use some philosophical language, it is not a comprehensive tract in any of these disciplines. I will not deal with the history or psychology of religious belief, nor the role of religion as a source of moral teachings. I will only peripherally mention religion's role in providing for personal emotional needs and not go deeply into the secular alternatives. All of these are amply covered by other books.

I also will not deal with religion as a social phenomenon, with one important exception. As already indicated, society, primarily in the United States, is now presented with a major conflict between science and religion—fueled by the rhetoric of those who see science as a threat to traditional faith. As mentioned, this conflict is being fought almost exclusively in the public and political arenas and not on the pages of scientific journals, nor at the thousands of international meetings mainstream scientists hold yearly. Still, its impact on the issues discussed in this book is so great that it cannot be ignored.

I will not consider in any detail the philosophical and theological disputes on the existence of God and the supernatural that have continued from ancient times to the present—the "rational" and "logical" elements of *logos* put to use in the absence or ignorance of science. Again, many books have covered these disputes, and I see no need to say yet again what has already been said thousands of times in thousands of ways without reaching a definitive conclusion.

My arguments will rest on the three elements of *logos*: reason and logic, but most important, science. Because it bases its judgments on empirical facts and not clever rhetoric, scientific method sports a track record of success unmatched by any other discipline. This has earned special recognition for science as the prime tool for arriving at definitive conclusions about reality. To avoid being hypocritical, those who would deny science this role had better do so on stone tablets rather than the printed page, and via smoke signals rather than the Internet.

As I will attempt to show, the empirical data and theories based on that data are now sufficient to make a scientific judgment: In high probability, a nonmaterial element of the universe exerting powerful control over events does not exist.

The scientific questions I will discuss divide into three types: (1) those that deal with empirically based theories of the origin and nature of the universe and its laws; (2) those that deal with the empirically based theories of the origin and nature of life; and (3) those that deal with direct empirical claims for God or the supernatural. Typical is the 1997 book *God: The Evidence*, by Patrick Glynn.[16] Glynn writes as if he is totally unaware of the scholarly refutations that already existed at that writing for many of the arguments he makes.[17] Currently, these topics, as they affect religious ideas, are being addressed almost exclusively in the public forum. There the discussion is marked by considerable confusion and contention, resulting from the lack of scientific sophistication of the participants. One would prefer to see these questions raised more frequently in academic circles, where scientific understanding is greater and reason has a somewhat better chance of prevailing. Unfortunately, this has not yet happened on a significant scale in secular institutions, where the subject is often deliberately avoided, in order not to offend sources of funding.

I will spend considerable time on (1) the origin and nature of the universe and its laws, because here is where my greatest expertise lies and where I feel that I have a number of original contributions to make. While I will rely strictly on existing, well-established theories of physics and cosmology, the philosophical implications of the latest developments in these fields have not yet been fully explored, even in academic circles. I will try to clear a path to the new territory being opened up.

Most scientists, philosophers, and theologians, as well as other academics and the general public, are unaware of the new insights on the origins of the cosmos and natural law that have come from recent developments in physics and cosmology. In particular, I will show that a scenario of a purely material, self-contained, uncreated universe can now be drawn that is consistent with, and is indeed strongly suggested by, existing knowledge in physics and cosmology.

I have less expertise on (2) the origin and nature of life, at least as far as detailed biological or paleontological matters are concerned. This is of no great import since these subjects are also already covered by a huge literature. Nevertheless, I think I have something new to contribute on this subject as well, in light of the recent insertion of information theory into the debate over the evidence for design in the universe. I will discuss the connection between information as defined by information theorists and entropy as defined by physicists and show how some of the assertions made by design theorists conflict with well-established principles in both fields. While design theory, as currently formulated, has been widely, and properly, criticized on many other grounds, I will show that some of its most important arguments are not just probably wrong but provably wrong, making the claim that design theory is a legitimate new science highly dubious.

As an experimentalist, I feel I also have original contributions to make with respect to (3) the strictly empirical questions concerning the existence of a world beyond matter. The popular media frequently report that paranormal, supernatural, or spiritual effects have been observed and measured both in the field and in laboratories, and published in scientific journals. These range from anecdotal stories about ghosts and miracles, to epidemiological studies that purport to prove the efficacy of prayer, to claims that psychic phenomena have been solidly confirmed by experiment. While the demonstration of psychic phenomena might not imply anything supernatural, this would become a popular explanation should the reports prove to be valid. I will show, however, that none of these reports come close to meeting the standards that science applies to any extraordinary claim.

I will also consider other claims that may appear, superficially, to relate only marginally to the question of the existence of God. Some of the most egregious are made by the proponents of alternative medicine. I will show that, despite what the media report on this subject, nontrivial alternative therapies that contradict existing science have not been validated by controlled clinical studies. If any were, they would already be part of conventional scientific medicine. And, as was the case for psychic phenomena, should any alternative therapy be validated in the future, a perfectly natural explanation might still be found.

However, the theoretical bases for many of the claims of alternative medicine, especially those that exploit the notion of "human energy fields," are already in gross conflict with much of established physics. If these therapies really worked for the professed reasons, then large parts of physical and biological science that already work well in other areas would have to be discarded. This, in turn, would leave a huge explanatory gap that, in principle, could be filled by the traditional notions of spirit on which most of these therapies are ultimately based. The implications of the demonstrated reality of psychic phenomena would be similarly revolutionary, supporting many ancient, prescientific intuitions about a reality that contains much more than matter alone and putting much of current physics into question. I will show that physics need not worry that it will be put out of business by alternative medicine or parapsychology.

I will not delve deeply into recent results in the fields of neuroscience and cognitive theory and the strong indications they are providing for a fully embodied mind, that is, a mind totally based on the material brain. Much of this is very new and still controversial, so I will limit myself to a physicist's gaze at the empirical data that are presented to support the opposing notion of a disembodied mind. Now, some critics may say that I am dodging the subject here. However, I do not feel I have the burden of proving the validity of the data supporting a fully embodied mind. This is the more parsimonious hypothesis, and the burden is with those who claim evidence for a disembodied mind. I will consider those claims and show that the empirical evidence is nil.

I do not fully accept Armstrong's characterization of *mythos*, quoted earlier, that it cannot be validated by rational proof. I take this to mean that *mythos* cannot be analyzed by any of the elements of *logos*. I suggest it can, by scientific means, and that the absence of such a validation makes it very likely that *mythos* provides no information about reality except the reality that exists solely inside a mystic's head.

The prescientific intuitions of transcendence and disembodiment that I will analyze are superstitious memories from the primitive past of humanity, based on *mythos*, that have been kept alive and given authority in the teachings of the world's great religions. These insist on a picture of reality in which humans are connected to the cosmos by supernatural bonds. Creation stories in scriptures tell us that we were placed on Earth for a divine purpose. Other scriptural tales tell us that the creator acts regularly to see that his purpose is fulfilled. While these narratives are generally regarded as the independent province of religion, most believers do not regard them as simple myths whose only reality is "embodied in cult, rituals, and ceremonies." Either literally or metaphorically, in the believer's mind religious narratives evoke an ultimate reality behind cosmic and earthly events. If this is indeed the true reality, then the existence of such a creator is amenable to scientific discovery.

Enlightenment thinkers, such as Thomas Jefferson and many of the other founders of the American republic, imagined a *deist* god—a perfect deity who created the universe and its laws that require no further intervention. Other thinkers, such as philosopher Baruch Spinoza and modern scientists such as Albert Einstein, saw a *pantheist* god—an impersonal god equated to the order of nature. These gods do not lend themselves readily to scientific study, although the Enlightenment deist god seems unlikely in the light of the quantum uncertainties of modern physics. In any case, neither the deist nor the pantheist god corresponds to the *theist* God who is worshipped by Christians, Jews, and Muslims. Their God is a personal one who plays an active, indeed dominant, role in the universe. In this book I will use a capital G for the theist God and lower case g for any other gods.

Some theistic scientists, such as those at the Berkeley meeting, say that the existence of a personal God is supported, even *required*, by current scientific knowledge. While I will dispute that claim, they and I at least agree that the existence of such a God can be established, to a high degree of certainty—by scientific means. That being the case, theists can hardly then disagree with the corollary to this statement, that the existence of a personal God can be ruled out, to a high degree of certainty—by scientific means. In this book, I argue that, based on the accumulated data, the latter conclusion is in fact the more rational one.

In making my case, I will find it necessary, on a few occasions, to get a bit technical. These sections should be understandable to anyone just slightly familiar with elementary scientific and mathematical methods. This enables me to state my arguments more precisely and quantitatively, to show that I am presenting not vague opinions, but concrete conclusions that have a sound empirical and theoretical basis. For those who cannot follow these details, I will also try to summarize the argument in more familiar terms, so the reader is encouraged to keep reading. Three technical appendices are also included that require a knowledge of physics at about the freshman or sophomore university level.

I do not expect that many believers will change their beliefs based on the scientific arguments I will present here. They may have other reasons to believe, ones based on *mythos* rather than *logos*. My only goal for any from that group who happen to pick up this book is to help them to realize that their beliefs have no scientific foundation, no *logos*. They should not attempt to convert others to their beliefs on the basis of the highly dubious arguments they may hear, which claim to be science and are often, in fact, downright falsehoods. Faith based on science will quickly turn to nonfaith when the science becomes better understood. My advice to the adamant believer is to stick to *mythos* to justify that belief.

Any strategy that attempts to reinforce faith by undermining science is also doomed to failure. Showing that some scientific theory is wrong will not prove that the religious alternative is correct by default. When the Sun was shown not to be the center of the universe, as Copernicus had proposed, Earth was not moved back to that singular position in the cosmos. If Darwinian evolution is proved wrong, biologists will not develop a new theory based on the hypothesis that each species was created separately by God 6,000 years ago.

Some will say I am preaching to the choir. However, choirs, too, can benefit from some preaching. Nonbelievers should find their intuitive nonbelief reinforced by rational, scientific arguments and hopefully gain a greater understanding of the strong scientific foundation that undergirds their nonbelief. At the very least, they will be better able to respond knowledgeably to those theists who challenge them to justify the materialist model of reality. That challenge is not to be impatiently dismissed as "nonsense," but to be answered thoughtfully and respectfully.

Unlike the general public in the United States and many other countries, the majority of scientists are nonbelievers. And, despite what *Newsweek* says, the number who say they see evidence for God in science is not growing. If anything, surveys indicate that belief among scientists is falling off. Nevertheless, most scientists would probably classify themselves as *agnostics* rather than *atheists*. That is, they would say that they simply do not know if God exists. I hope some of these agnostic scientists will take a more careful look at the empirical data and realize, as I have, that these data are already sufficient to make a strong, scientific statement about the very likely nonexistence of the Judeo-Christian-Islamic God. I also hope that scientists will realize that they cannot sit back and ignore the current challenges to science being made by religious zealots who wish to suppress this fact.

I hope that agnostics will see the merit of taking a stronger position of disbelief. These would include both academics and laypeople who will not have heard many of the arguments to be presented here. I hope that nonscientist academics and laypeople will learn that science has a strong case to make for a purely material reality, one that is more than just the prejudice of a particular culture of "scientism."

Finally, I hope the reader will not come away thinking that I am just a cold, bitter materialist nihilist who sees no meaning in life. In fact, I am a happy man who has found much meaning in my life, in work and family, in art and music, in sports and travel, and, indeed, in almost everything the world has to offer, except fantasy and mythology. Science is a big part of my life, but not the only part. Someday I may write about some of the other parts, but here I will stick to the science.

Considerable material supplementary to this book and links to even more can be found on my Web site, currently located at http://spot.colorado.edu/~vstenger/. The reader is invited to join in the e-mail discussion group (avoid-L@Hawaii.edu) that operates from this base.

Some eighty people, from a wide range of backgrounds, not just science, helped in the preparation of this manuscript. As I did with my two previous books, I placed draft chapters on the Web and called for comments that were provided by means of the e-mail group. The comments ranged from small corrections to grammar to major issues that produced considerable debate. I am deeply grateful for all the help I received, large and small. The following long list gives the names of those who provided me with useful information, made comments on the manuscript, or participated in the discussions—usually all three: Ivan Antonowitz, Dean Batha, Greg Bart, Dan Brennan, Jean Bricmont, Richard Carrier, John Christoper, Jonathan Colvin, Kevin Courcey, Liz Craig, Patrick Curry, Ermanno D'Annunzio, Lloyd Davidson, Keith Douglas, Ron Ebert, Taner Edis, Wesley Elsberry, Peter Fimmel, Michael Fisher, Yonathan Fishman, Victor Gijsbers, Paul Gross, William Harwood, Ian Hill, Burt Humburg, Jim Humphreys, Stanley Jeffers, Bill Jefferys, Larry Johnson, Pat Johnson, William Keener, Ludwig Krippahl, Les Lane, Norman Levitt, Justin Lloyd, Richard Lubbock, John Mazetier, Don McGee, Scott McGlasson, Brent Meeker, Keith Miller, James Milstein, Clayton Naff, Raymond Nelke, Edward Olech, Anne O'Reilly, Mark Perakh, Bob Phillipoff, Norman Phillips, Markus Poessel, Jorma Raety, Todd Rockhold, Jason Rodenhouse, Linda Rosa, Paul Rybski, Chris Savage, Terry Savage, Thomas Schneider, Claus Segebarth, Jeffery Shallit, Peter Smitt, Wayne Spencer, Lester Stacey, Greg Stone, John Stone, John Syriatowicz, Erik Tellgren, Phil Thrift, Farrell Till, Chester Twarog, Dan Vergano, Cliff Walker, Richard Wein, Edmund Weinmann, William Westmiller, Kim Whitesell, Thair Witmer, Jim Wyman, and Roahn Wynar. Of course, none is responsible for any errors that may remain and none should be assumed to endorse anything within. Several sharply disagreed with me, and I welcomed the chance they provided for me to correct or reformulate my arguments. I believe this book is far better than it might have been had I not had the benefit of all this help.

NOTES

1. Remarks on *700 Club*, September 13, 2001 [online], http://www.truthorfiction.com/rumors/falwell-robertson-wtc.htm.

2. Anne Graham Lotz, interview by Jane Clayson on *CBS Early Morning* [online], http://www.cbsnews.com/earlyshow/healthwatch/healthnews/20010913terror_spiritual.shtml.

3. "Faith and Doubt at Ground Zero," *Frontline* [online], http://www/pbs.org/wgbh/pages/frontline/shows/faith.

4. Sharon Begley, "Science Finds God," *Newsweek*, July 20, 1998.

5. George Johnson, "Science and Religion: Bridging the Great Divide," *New York Times*, June 30, 1998.

6. George Johnson, "Science and Religion Cross Their Lines in the Sand," *New York Times Week in Review*, July 12, 1998.

7. Gregg Easterbrook, "The New Convergence," *Wired*, December 2001, pp. 165–69.

8. Center for Theology and the Natural Sciences [online], www.ctns.org.

9. Templeton Foundation [online], www.templeton.org.

10. Discovery Institute [online], www.discovery.org. Follow links to the Center for Science and Culture, where its mission statement, list of fellows, and other information can be found. Note that this center was formerly named the "Center for the Renewal of Science and Culture."

11. The CSC "wedge strategy" currently can be found online at http://www.public.asu.edu/~jmlynch/idt/wedge.html. This has been removed from the CSC Web site, but its authenticity is documented. See James Still, "The Wedge Strategy Three Years Later" [online], http://www.secweb.org/asset.asp?AssetID=200.

12. Karen Armstrong, *The Battle for God* (New York: Ballantine Books, 2000), p. xv.

13. Ibid.

14. Ibid., p. 376.

15. Ibid., p. 135.

16. Patrick Glynn, *God: The Evidence* (Rocklin, Calif.: Prima Publishing, 1997).

17. Michael Martin, "Review of Patrick Glynn's *God: The Evidence*" [online], http://www.infidels.org/library/modern/michael_martin/glynn.html.

ONE

THE BELLS OF
ST. MARY'S

Man is an imperceptible atom always trying to become one with God.
—Henry Adams, 1904

In the spring of 1941, as world war loomed, I was in second grade at Public School Number 12 in Bayonne, New Jersey. For several weeks I was dismissed early so I could attend Catechism at nearby St. Mary's Catholic Church. My parents were Catholic, but they sent me to public school, despite being told in no uncertain terms by the priests and nuns at St. Mary's that they were placing themselves in danger of eternal damnation.

My father was a Lithuanian immigrant, and my mother's parents had been born in Hungary. The Lithuanian church, St. Michael's, was too small to have a school, and there was no Hungarian church in a town where religion along with everything else was partitioned by ethnicity. We were outsiders in a working-class, immigrant neighborhood that was mostly Irish and Polish and as Roman Catholic as only the Irish and Polish can be. St. Mary's was an Irish church, and I guess my parents did not think I would fit in. Later, in my teens, I found the Irish priests rather delightful when my newspaper route took me into the various bars where they hung out for free drinks. The nuns were scary, though, to a six-year-old—far more forbidding than the Protestant spinsters who taught in the public schools.

Those were the days when essentially only five respectable careers were open to women outside the convent: mother, secretary, nurse, librarian, or school teacher. Many of the most intelligent women became school teachers, so even public schools in poor neighborhoods were excellent. I would never have made it into the academic world without the start those wonderful women gave me.

Nevertheless, off to St. Mary's it had to be for my preparation for First Communion. I still remember the nuns, with their indistinguishable bodies covered in black and fierce faces framed in white, drilling us in the Catholic Catechism. Q: "Who made the world?" A: "God made the world." I have a bizarre memory of their explanation of the differences between mortal and venial sin: When you are free from sin, such as just after confession, your soul is white. When you commit a venial sin, such as eating meat on a Friday, your soul turns gray. And, if you commit a mortal sin, such as that most horrible one of all, missing mass on Sunday, your soul turns black!

Now I had no idea what a soul was. But my family had the usual pious pictures of Jesus and Mary on the wall, with light radiating from their sacred hearts. So it was natural for me to identify the soul with the heart, and I imagined my heart turning black when I missed mass.

The nuns also taught us the Our Father and Hail Mary, which I had already heard at home. I always recited the Hail Mary:

Hail Mary, full of Grace, the Lord is with thee. Blessed art thou among women, and blessed is the fruit of thy *whom* Jesus.

It never made any sense to me. Why was Mary full of my Aunt Grace, who lived on Staten Island? And no one ever corrected me by explaining that "thy whom" was supposed to be "thy womb." That would have required them to tell me what a "womb" was, which was far too sexy for a six-year-old. So, "thy whom" it remained.

The most dreaded event that we prepared for was not First Communion but First Confession on the day before. I did not sleep well in the week prior to that awesome event. Then the fateful Saturday arrived. We all lined up outside the confessionals in the spooky, candle-lit chapel in the basement below the main church. Finally my turn came. I kneeled inside the even darker booth until the invisible priest pushed back the sliding window that separated me from his chamber. "Bless me Father, for I have sinned. This is my First Confession. I have lied ten times, disobeyed my parents five times, missed mass two times." The bored priest responded: "Say five Our Fathers and ten Hail Marys" (or whatever), and it was all over.

It felt as if a great weight had been lifted from my head as I moved to the altar to do my penance. I walked away from the church with an immense feeling of serenity and harmony with God. I can understand, from my memory of visits to the confessional, how people have mystical experiences that they associate with great powers beyond themselves and how rituals can trigger these experiences.

My peace after First Confession, however, was short-lived. Several classmates who had been behind me in the confessional line caught up and started to tease me. I had spoken too loud (something I generally do, being deaf in one ear), and they had heard my every word in the confessional. "I have lied ten times, disobeyed my parents five times, missed mass two times, hah, hah, hah." How mortifying!

When I was growing up, Bayonne was an oil-refining town just across the bay from Manhattan, near the Statue of Liberty. It had about 70,000 people, of which I would guess over 50,000 were first- or second-generation immigrant Catholics from eastern Europe and somewhat earlier-generation Catholics from Ireland and Italy. The remainder of Bayonne's population was comprised of a few thousand well-educated Jews, who owned almost all the businesses and were most of the doctors, lawyers, and dentists. Also included were a handful of Protestants, who were largely factory workers like the Catholics or had somewhat better jobs such as office work or teaching in the public schools. There were also a few thousand "negroes," who were regarded as another species though I often played ball in the parks with them.

Religion was an obsession among Bayonne Catholics. I never ran into an evangelical or fundamentalist Protestant. Nevertheless, no one would ever admit to being an atheist. At least, I never heard of anyone doing so.

Actually, my P.S. 12 teachers were about the only Protestants I met, aside from my mother's friend Mrs. Rehill and her family. There were a few Jewish families who lived a half-block away on the corner of Avenue A and Tenth Street and ran the corner grocery and candy stores. I played with Leonard Goldberg. Once he invited me to the Jewish Community Center where some of the stars from my favorite Saturday morning radio program, *Let's Pretend*, were doing a show for kids. I entered the hall with great trepidation, expecting to be struck down by lightning. The nuns had warned us never to enter a Jewish synagogue or Protestant church on pain of Hell. Since the hall was not a synagogue, my mother figured it was okay and let me go.

Religion was supposed to be a comfort to the working people of Bayonne, whose lives were hard, though far from oppressive. They had experienced the Depression, World War II, and then the threat of communism and nuclear holocaust, all while I was growing up. Brothers and husbands had died in war or suffered injury. Parents aged quickly and spent their last days in pain.

Two of my older cousins died as children, and a playmate suddenly succumbed to polio. A child next door was killed by a car on Avenue A. The parents in these tragedies grieved to the point of madness and never truly recovered. I do not think their religion gave them any peace. In fact, I am convinced that a great portion of their intense pain was the result of guilt brought upon them as a direct result of their faith. Although their well-meaning and sympathetic priests assured them it was not their fault but simply God's will, the parents spent the rest of their lives agonizing over why God would have done this to them except as a punishment for their sins. They might have been able to handle it better had they been nihilists and simply regarded their tragedies as random events in a meaningless cosmos. Then, after recovering from their initial grief, they might have gone on with more normal lives. But I know what a difficult position that is to take.

Most people in Bayonne, like folks in similar towns across the country, had little education and could neither verbalize nor intellectualize their problems very well. They just suffered them. They listened eagerly when the priests promised them everlasting life in paradise, where they would be reunited with their departed loved ones, but this was not enough when the suffering and guilt were unbearable.

The parish priests did their best, and I fault them little. They operated within a framework developed over centuries that would not have survived this long if it did not give people something they wanted, no matter how insufficient. Many years later when my father lay dying, I rushed to Bayonne from England, where I had been spending a sabbatical at Oxford. After my father's funeral service at St. Michael's, my brother and I were escorting our mother down the aisle when we were met halfway down by the priest who had performed the ceremony. Mother had not participated in the communion and in front of the whole congregation explained to the priest: "Father, I just did not have time to go to confession." The priest looked at her with great compassion, made the sign of the cross in front of her eyes, and said simply, "Your sins are forgiven." Even though by that time I regarded the existence of God as very unlikely, I still felt better. I know Mom did. Brother Bill, by then a devoted member of Herbert Armstrong's Worldwide Church of God, kept his thoughts to himself.

As I grew up in Bayonne, I increasingly disagreed with my Catholic friends and family. Although memories can be quite inaccurate, I recall that this happened as the result of my own reading and thinking, without the influence of a single person with whom I talked face-to-face, heard on the radio, or saw on television. I do not recall meeting anyone—no teacher and certainly no family member—who was not a believer, or, at least, admitted nonbelief. I do not recall reading any atheist or agnostic literature. This was as hard to find as pornography, and I found little of that, either.

My mother became increasingly religious as she got older. My father rarely went to church, but he remained a Catholic and always expressed belief in God. He did not argue with me about my views—although he and other older relatives often told me to keep my mouth shut. They said I had no business questioning the authority of the Church. While they succeeded in keeping me from expressing my thoughts too openly, they had no effect on those thoughts. As long as I kept my mouth shut, they left me alone.

I now understand my father's position, although I did not at the time. He had grown up fighting on the tough streets and was very familiar with the political corruption of the Hudson County machine and the activities of organized crime. He had not prospered from this knowledge, because he was an honest and moral man, and he avoided trouble by keeping his mouth shut.

As I learned later when I lived in Hawaii, the Japanese have an expression that describes this cautionary attitude that they drill into their children: "The nail that sticks out gets hammered back down." But I was not subject to the same pressures as my father and quickly learned the opposing New Jersey credo: "The squeaky wheel gets the grease."

Like most of the immigrants of his generation, my father was just glad to be in America. He usually said he was born in Brooklyn, not Lithuania, and never dreamed of going back, even for a visit. "God bless America!" was the common toast heard at family holiday dinners. Despite the heavy hand of authority hanging over them from the Church, corrupt officials, and the factory floor, the immigrants and their children were still far freer than those left behind in the Old Country. They were grateful for the few handouts and other favors that the political machine arranged to ensure their votes. They did not mind being told what to do by the Church, either, if it would guarantee for them a better life beyond the grave. And they certainly were not going to buck the Mafia.

The Catholic Church's authority had come down in an unbroken line of popes from Peter, according to what the priests told their parishioners. They never mentioned Pope Sixtus IV, for whom the Sistine Chapel is named, earning for himself an extra income of 30,000 ducats a year from the brothels of Rome. Neither did they talk about Pope Alexander VI, who had ten illegitimate children and surely broke some kind of record: sleeping with the women of three generations—his mother, his mistress, and his daughter (the infamous Lucrezia Borgia).[1]

Whenever I brought up such unpleasant facts or questioned Church teachings, I was told, in effect: Who are you to think you are smarter than the priests, bishops, pope, and almost two thousand years of revelation? What kind of world would this be if everybody decided for himself what was right or wrong?

It was not that I thought I was smarter. I had simply explored science and found what seemed to me a far more powerful authority. And I did not steal or murder because I thought they were wrong, not because I feared damnation.

I had become fascinated with science after World War II, particularly the nuclear bomb. I recall listening to the Bikini atoll tests on the radio, and I read every book I could find in the library on what was then called "atomic energy." To have such awesome power, science had to possess something special.

By contrast, I saw little power in the religion all around me. The priests comforted the sick, but the doctors cured them. And when the doctors could not, all that praying and blessing did not change the outcome.

The power of science was evident everywhere, and none of it came from the Bible or other sacred texts. In fact, science seemed to have developed in opposition to religion and to have triumphed despite this formidable opposition.

Of course, I am expressing these thoughts from my current,

comparatively sophisticated perspective decades later. I certainly did not have these ideas clearly formulated in my thoughts back then. But, jumbled and dimly perceived as it was, my thinking steadily flowed like solar particles toward a scientific pole that was clearly opposite to the pole of religious thinking toward which the thoughts of those around me flowed.

A visit to the Hayden Planetarium in Manhattan only accelerated this flow, as I added astronomy to my reading list. I was enthralled by the paintings of the planets and moons by Chesley Bonestall, imagined myself cruising the solar system, and worried that I would be too old to be first to fly to the moon. It turned out I was too young.

I learned that Earth was only a small planet circling one of billions of stars in one of billions of galaxies. This struck me then, and it still strikes me now, as the very antithesis of the biblical view of Earth residing at the center of a universe, created by God as the home of his special creation, formed in his image, Man.

This conclusion was further reinforced in my readings on biological evolution. I remember thinking how much sense it made, and excitedly telling others about it. I assumed they would see the sense, too, once I pointed it out. But they did not.

My library searches had even uncovered a book by a Catholic priest, complete with Church imprimatur, that explained how the Church did not forbid belief in evolution. By that time the influence of Jesuit paleontologist Pierre Teilhard de Chardin had been felt, and Catholic theologians were beginning to view evolution as one more piece of evidence for design in the universe, with Man at the pinnacle of natural progress following the laws set down by God at the Creation. I tried to explain this to those around me, but no one listened. I suppose that was just as well, because this (teleological) interpretation of evolution assumed humans to be evolving toward some predetermined state, which is not at all what the theory of evolution implies. As I discovered much later, biological organisms exhibit many qualities that indicate a largely random process of evolution, guided only by natural selection and the limitations of physical law.

As I learned more about physics, it began to seem reasonable to me that an all-perfect God would have created the universe complete with physical laws that would lead to his desired ends without further intervention. Such intervention would be unnecessary since the laws would, by the nature of God, have to be perfect to begin with. The galaxies, stars, planets, and life would all develop naturally, according to God's plan. Later I learned that this was the view of the deists of the Enlightenment.

The big bang theory had not yet been developed in those days, but it would later become a major element in the teleological-theological scheme that I had been gradually absorbing from my reading. In 1951 Pope Pius XII offered the big bang as a "proof" of *creation ex nihilo* and the existence of a creator. This, along with sophisticated arguments about how the universe seems to be fine-tuned for life, forms part of a modern version of the argument from design that we will consider in detail later.

The natural theology of a world machine operating according to God's laws, inexorably moving toward the ultimate fulfillment of his plan, had developed from the time of René Descartes and Isaac Newton. It was fortified by interpretations of the discoveries of Charles Darwin and Albert Einstein to become, in generic terms, the common view still held by scientists and those theologians who have adapted scientific perspectives to suit their needs. While their image of God ranges from the personal savior of Christianity to the impersonal order of nature, many scientists and theologians today see the scientific search for a *theory of everything* (TOE) as an attempt at reading the "mind of God."

By the time I attended the Newark College of Engineering (now New Jersey Institute of Technology), I had stopped going to church. Later, while in graduate school at the University of California at Los Angeles (UCLA), I started attending the Westwood Community Methodist Church on Wilshire Boulevard. With its millionaires and celebrity parishioners (I once shook Richard Nixon's hand, after an Easter Sunday service), it was a far cry from St. Mary's. I had some interesting theological discussions with the liberal, sophisticated ministers. But my interest in Westwood was not theological but rather singing in its spectacular choir and meeting girls. Indeed, I met my wife at Westwood.

After marriage and my doctorate in physics, my wife and I moved to Hawaii where we resided and raised two children until our recent relocation to Colorado. Our lives were enjoyably interrupted by sabbaticals at the University of Heidelberg in Germany, Oxford in England (twice), and a national laboratory in Italy. We spent many summers elsewhere, in delightful places ranging from Berkeley, California, to Aspen, Colorado, and Florence, Italy. We were in Russia when communism was falling; we were in Berlin when the Wall was coming down, and have souvenir pieces that are probably worth something today.

During these years I participated in the most exciting period of the development of elementary particle physics. This was an era of abundant government funding for science, motivated by the Cold War, and I happened along at the right time. I collaborated in some of the experiments that made important discoveries, such as charmed quarks and gluons, and had a major role in the development of high-energy neutrino astronomy. In my last research project, I was one of the 120 physicists who produced strong evidence that the neutrino has mass.

I mention this, not to toot my own horn, but to establish my credentials as a legitimate, mainstream scientist who is very familiar with the process by which new, fundamental knowledge about the nature and structure of the universe is gained. I was there, working and watching from close-up, when all the important discoveries in physics and astronomy of the second half of the twentieth century were being made. I first heard Richard Feynman lecture on quantum mechanics in 1956. I have talked physics, one-on-one, with Feynman, John Bell, Steven Weinberg, and other celebrated physicists of the era. I have heard many others lecture, and I have read their papers and books.

From this experience, I have learned what science asks of us when we claim the existence of an extraordinary new phenomenon. It requires much, including years of hard work, uncompromising honesty, and willingness to accept failure. I can quickly recognize fallacious logic or faulty experimental procedure when I read a paper that purports to observe something that goes beyond existing knowledge. I am dubious and suspicious whenever an important result has been obtained too easily or too quickly, and reported in the media before it has run the gamut of critical review by disinterested, knowledgeable parties.

By the time I reached my fiftieth year, in 1985, I started looking beyond the teaching and research that had occupied me full time until then and returned to the more philosophical concerns that first caught my attention as a youth. I started a second career, writing and speaking on subjects where my knowledge of physics and astronomy intersected with philosophy, theology, and in areas where claims were being made on the fringes of science.

I began to publish on these subjects in 1988, with *Not by Design: The Origin of the Universe*. In that book I attempted to show that no principle of physics was violated by a universe that appears from nothing. Neither the first law of thermodynamics (conservation of energy), nor the second law (an isolated system left to itself will not become more orderly on the whole). I tried to explain the basic physics and cosmology for the general reader and described the latest models. In the present book, I will revisit and extensively update these ideas and explore the new territory that has come to light during the recent rapid development of the science-religion dialogue.

Since my goal has always been to understand the fundamental nature of the universe, I also began to probe the connection between physics and paranormal phenomena. I discovered the large body of popular literature that promoted psychic claims. Many authors were asserting that extrasensory perception (ESP), mind over matter, and other wonders were supported by scientific evidence. The dramatic implication was that a scientific basis existed for the reality of spiritual forces.

This led to the 1990 publication of my second book, *Physics and Psychics: The Search for a World Beyond the Senses*. There I examined the evidence for psychic phenomena from my particular physics perspective. I found that the data did not support the extravagant claims being made for psychic phenomena. While much of the work in this field was being done by physicists, no more than two or three of these reports had been published in reputable journals such as *Nature* or the *Physical Review*. Furthermore, nothing published anywhere came close to meeting the kind of stringent criteria my colleagues and I were used to seeing applied by these journals to our own published work.

I was astonished to discover that quantum mechanics was being widely used by promoters of paranormal claims to provide a supposed theoretical justification for the mind's ability to communicate with other minds, move objects, and, in effect, control reality. Lately, this dubious quantum connection has been picked up by those who promote nonscientific alternative medicine. My mainstream physics colleagues and I reacted in dismay when we heard these bizarre assertions, often made by physicists. My response was to write *The Unconscious Quantum: Metaphysics in Modern Physics and Cosmology*, which appeared in 1995.

This book was not intended as a polemic against ideas that were outside the consensus of physics. I probed deeply into the philosophical disputes over the interpretation of quantum mechanics that, like most physicists, I had hardly known existed. To us, quantum theory was undisputed. Paranormalists, however, saw an opening for their mystical beliefs in the lack of agreement among experts on what some of the strange behavior observed in quantum experiments implied about underlying reality and the role of human consciousness.

While my intense study of this matter did not lead me to change my opinion that quantum mechanics was being grossly misused by the paranormalists, I discovered that the quantum offers a rich source of profound speculation about the nature of reality that was not available to philosophers and theologians of previous centuries. Indeed, I began to wonder how any philosopher writing before 1925 could have said much at all about reality, not knowing about quantum mechanics or, equally significant, Einstein's theories of relativity.

I began to see that my many years spent in particle physics

probably positioned me closer to glimpsing the nature of reality than any pursuit of philosophy could have done. That is not to denigrate the brilliance of the philosophers of previous centuries; they simply lacked an important piece of reality's puzzle. Thought, without the data on which to structure that thought, leads nowhere.

Reality is what kicks back when you kick it. This is just what physicists do with their particle accelerators. We kick reality and feel it kick back. From the intensity and duration of thousands of those kicks over many years, we have formed a coherent theory of matter and forces, called the *standard model*, that currently agrees with all observations.

I described the picture of reality that is implied by these discoveries in my next book, *Timeless Reality: Symmetry, Simplicity, and Multiple Universes*, published in 2000. Basically I argued that the model of reality suggested to us by modern physics and cosmology is a modernized version of the conceptually simple and purely material one proposed by Democritus and Lucretius over two millennia ago. In this model, localized fundamental bodies move around in an otherwise empty void, interacting with one another only when coming in contact or exchanging other bodies.

Furthermore, the space-time manifold that is used to describe the behavior of these bodies requires no special origin or direction. This fact alone is sufficient to give our universe most of its basic, global laws: the conservation principles of energy, momentum, and angular momentum, along with the results of Einstein's theories of relativity. In particular, the absence of a preferred direction or "arrow" of time at the fundamental level is strongly implied by quantum experiments. The refusal of many physicists to accept this fact and to force the psychological time arrow of everyday human experience into their descriptions of quantum experiments leads to the appearance of the so-called quantum paradoxes and the unnecessary introduction of bizarre and mystical elements into quantum mechanics that give such aid and comfort to paranormalists.

The phrase "timeless reality" calls into question some of the most basic notions that appear in theology and metaphysics. No beginning of time implies no creation. No special direction of time implies no distinguishable cause and effect. In *Has Science Found God?* I will demonstrate how powerfully these ideas impact philosophy and theology and attempt to show why much recent as well as ancient thinking about the nature of reality has to be greatly modified in the light of the developments of modern physics and cosmology.

NOTE

1. Peter De Rosa, *Vicars of Christ: The Dark Side of the Papacy* (London: Corgi Books, 1988).

TWO

THE MENACE OF DARWINISM

The universe we observe has precisely the properties we should expect if there is, at bottom, no design, no purpose, no evil and no good, nothing but blind pitiless indifference.

—Richard Dawkins, 1995

THE DANGER OF THE SIMPLE

The popular perception of the history of science is an imagined series of revolutions in which old theories are abruptly cast on the dustheap as new theories take their place. This scenario was bolstered by physicist-philosopher Thomas Kuhn, whose widely read *The Structure of Scientific Revolutions* introduced the now familiar term *paradigm shift* to our common vocabulary.[1] However, the notion that science proceeds by abrupt transitions is, at best, an exaggeration. Nobel laureate–physicist Steven Weinberg maintains that the history of science shows only a few examples of true Kuhnian mega-paradigm shifts.[2] In fact, most of the changes that take place in science are gradual, with old theories often remaining in use long after new ones have come along. New theories tend to expand into new territory rather than take over the old.

For example, classical Newtonian physics still constitutes the bulk of the physics curriculum and remains in wide use a century after the quantum revolution supposedly showed that it was "wrong." Weinberg notes that even Kuhn, who was his colleague at Harvard, taught classical physics to his students. Quantum mechanics encompasses the classical domain, but its unique applications occur in other domains, such as the atomic and subatomic, in which no data to check against theory existed at the time Newtonian mechanics was being developed.

Nevertheless, at least two scientific advances in the last half-millennium merit the label of major or mega-paradigm shifts: (1) Copernicus's sixteenth-century proposition of Earth's motion around the Sun, and (2) the nineteenth-century hypothesis of the evolution of species by natural selection put forth by Charles Darwin and Alfred Russel Wallace in 1859.

Both of these developments moved human thinking into new territory, but they also took over the old, displacing previously existing, deeply entrenched systems of thought. Most notably, they explicitly contradicted traditional beliefs based on holy scriptures that were believed to infallibly reveal the word of God. At the time it appeared, each proposal was seen as a mortal threat to the Christian faith. In this chapter we will see how Darwinism is still regarded by a vocal minority of Christian believers, mainly in the United States, to be such a deadly menace to their faith that it must be fought against by every possible means, fair or foul.

It should be kept in mind that modern evolutionary theory has advanced far beyond Darwin's original ideas, most notably in its discovery of the basic molecular mechanism involving DNA. Furthermore, other natural processes besides natural selection are thought by some to play a role. While definitions may differ, I will use the term Darwinism to refer to the evolution of biological organisms by the purely natural processes of chance and natural law.

The sixteenth-century Roman Church forced Galileo to disavow any implications from his writings that the Copernican theory of the solar system was a fact of reality and not simply a mathematical model, as Copernicus's publisher had carefully declared. Scripture is quite precise in stating that Earth is the immobile center of the universe: 1 Chron. 16:30 says, "Yea, the world stands firm, never to be moved." Ps. 104:5 affirms, "Thou didst set the Earth on its foundations, so that it should never be shaken." Similar references can be found in Ps. 93:1 and 96:10.[3]

Nevertheless, within a century or so after Galileo, theological resistance to the Copernican cosmology had largely dissipated. The supporting evidence was simply too overwhelming, and religion had to adapt or die. In a kind of natural selection, it adapted and sanctioned the notion that Earth, in reality, moves about the Sun. However, this forced churchmen to admit that what is written in the Bible cannot always be taken literally. Thus began the art of *apologetics* in which the writings in the Bible are reinterpreted so as to conform with new knowledge, no matter how contradictory that knowledge may be to the most straightforward reading of the words therein.

By the nineteenth century, ecclesiastical power had greatly diminished in Europe, and Darwin did not meet the same fate as Galileo. In fact, he ended his illustrious life interred alongside Newton in Westminster Abbey. (Galileo rests in a place of honor in Santa Croce cathedral in Florence.) Still, what philosopher Daniel Dennett has called "Darwin's dangerous idea," that humans and other living things evolved from less complex forms of life by purely natural processes unguided from above,[4] was hardly accepted with open arms at the time. Today, Darwin's dangerous idea remains the primary battleground in the war between science and fundamentalist religion.

In his monumental work, *A History of the Warfare of Science and Theology in Christendom*, published in 1896, Andrew Dickson White, the first president of Cornell University, reported the reactions of churchmen when Darwin's theory exploded on the scene. The following quotations, unless otherwise indicated, are taken from a 1993 reprint of White's work, which can be consulted for the original references.[5]

Bishop Samuel Wilberforce of Oxford (d. 1873) protested that "the principle of natural selection is absolutely incompatible with the word of God. [It] contradicts the revealed relations of creation to its Creator." Wilberforce, interestingly, uses an argument from parsimony similar to what we hear from modern theists, that there is "a simpler explanation of the presence of these strange forms among the works of God . . . the fall of Adam." Scientists also rely heavily on parsimony, and most see nature as the simpler alternative to God.

Wilberforce is notorious as the unlucky recipient of a zinger from "Darwin's bulldog," Thomas Huxley (d. 1895). During a confrontation at Oxford in 1860, Wilberforce, tongue-in-cheek, inquired of Huxley whether it was through his grandfather or grandmother that he claimed descent from a monkey. Huxley replied, "I would rather be the offspring of two apes than be a man and afraid to face the truth."[6]

White quotes an unnamed theological authority who lamented the depth of the implications of Darwin's idea: "If the Darwinian theory is true, Genesis is a lie, the whole framework of the book of life falls to pieces, and the revelation of God to man, as we Christians know it, is a delusion and a snare."

An unidentified representative of the American branch of the Anglican Church agreed with this dire assessment: "If this hypothesis be true, then the Bible is an unbearable fiction . . . then have Christians for nearly two thousand years been duped by a monstrous lie."

A certain Dr. Schund in Germany likewise viewed evolution as marking the death of Christianity, contending that "every idea of the Holy Scriptures, from the first to the last page, stands in diametrical opposition to the Darwinian theory. . . . If Darwin be right in his view of the development of man out of a brutal condition, then the Bible teaching in regard to man is utterly annihilated."

Following a pattern that continues to the present day, many books appeared that purported to "disprove" Darwin. In 1877, a French physician, Dr. Constain James, published *On Darwinism, or the Man-Ape*. Pope Pius IX (recently beatified on the road to sainthood) was delighted, writing to the author that he "refuted so well the aberrations of Darwinism. . . . [A system] which is repugnant at once to history, to the tradition of all people, to exact science, to observed facts, and even to reason itself, would seem to need no refutation."

It is interesting to note that the pope's objections were not stated in theological but in scientific terms, referring to "observed facts" and "reason." He continued in that vein, "But the corruption of this age, the machinations of the perverse, the danger of the simple, demand that such fancies, altogether absurd though they are, should—since they borrow the mask of science—be refuted by true science." The pope, in contradiction to Wilberforce, seemed to accept evolution as the simpler explanation but warns of "the danger of the simple." And he calls upon science, not theology, to refute Darwinism.

Not all theists reacted unfavorably to Darwinism. White relates that the "High Church party" at Keble College in Oxford (where I once had the privilege of dining at High Table) called evolution "an advance in our theological thinking." The bishop of London argued, "It seems something more majestic, more befitting him to whom a thousand years are as one day, thus to impress his will once and for all on his creation, and provide for all the countless varieties by his one original impress, than by special acts of creation to be perpetually modifying what he had previously made." And, despite the pronouncements of Pius IX, a statement from American Catholic sources declared that "the doctrine of evolution is no more in opposition to the doctrine of the Catholic Church than is the Copernican theory or that of Galileo."

By mid–twentieth century, the Catholic Church had largely accepted some aspects of evolution, or at least allowed Catholics to study it without pain of damnation. In a 1950 encyclical, *Humani Generis*, Pope Pius XII ruled that the "teaching Authority of the

Church does not forbid ... research and discussions ... with regard to the doctrine of evolution."[7] The Church position was substantially updated in 1996 by Pope John Paul II, who stated before the Pontifical Academy of Sciences:

> Today, almost half a century after the publication of the encyclical, new knowledge has led to the recognition in the theory of evolution of more than a hypothesis. It is indeed remarkable that this theory has been progressively accepted by researchers, following a series of discoveries in various fields of knowledge. The convergence, neither sought nor fabricated, of the results of work that was conducted independently is in itself a significant argument in favor of this theory.[8]

Some dispute exists on the proper translation of the original French, *"plus qu'une hypothèsis."* Did the pope say "more than a hypothesis," or did he mean that there was "more than one hypothesis"?[9] Reading the whole document, however, makes it clear that the pope was not disputing the scientific validity of the evolution of the material body. However, he strongly opposed any theory in which the mind is purely an evolved property of matter:

> Consequently, theories of evolution which, in accordance with the philosophies inspiring them, consider the mind as emerging from the forces of living matter, or as a mere epiphenomenon of this matter, are incompatible with the truth about man. Nor are they able to ground the dignity of the person.[10]

THE SCIENCE OF EVOLUTION

The battle over the validity of evolution has been publicly posed as a scientific one. However, you will find little sign of it in scientific journals, in which such quarrels as exist are over details, not the basic concept. Since Darwin's time, the empirical evidence in support of natural selection has multiplied manyfold. Evolution has proved so useful as a paradigm for the origin and structure of life that it constitutes the foundation of the sciences of biology and medicine.

We know so much more today than Darwin did, and what we have learned has deeply confirmed his essential intuitions and inferences. Most important, we now understand the fundamental ingredients of genetics and the role played by DNA. All of these developments have confirmed evolution's basic mechanism. Today, the intimate connections between all living things, their histories and common origin, can be read in their genomes. As I write, the results of the human genome project have just been published. The genome of every living organism that has been studied to date shows many common DNA sequences, providing strong evidence that all life arose from the same source—just as posited by Darwin. The evolution by natural selection of bacteria, fruit flies, and other organisms has been observed firsthand in the laboratory.[11] Evolution has proved an invaluable tool in medical research. Undoubtedly, many lives have already been saved because of the knowledge provided by evolutionary theory.

While, as we will see below, considerable debate exists on what criteria define an activity as being scientific, there can be little dispute that evolution is science. It deals with empirical observations and makes many testable predictions that have been confirmed. In particular, evolution by natural selection, as originally posed by Darwin and Wallace, was eminently falsifiable!

At the time of Darwin and Wallace, most people believed that the age of Earth was on the order of 6,000 years, as indicated in the Bible. Geologists were just beginning to gather evidence for a much older Earth, and this knowledge had a great influence on Darwin, who took Charles Lyell's classic *Principles of Geology* with him on the voyage of the *Beagle*.

In the first edition of *On the Origin of the Species by Means of Natural Selection*, Darwin made a crude estimate of Earth 's age, based on geology, of the order of several hundred million years. This, he reasoned, was sufficiently long for the processes of natural selection to take place and produce the wide range of species on Earth.

The great physicist William Thomson, later to become Lord Kelvin, disputed Darwin's estimate, arguing that Earth had a much lower age. Thomson had made major contributions to thermodynamics, formulating the second law of thermodynamics and establishing the absolute temperature (Kelvin) scale. At the time, the only known sources of energy that could account for solar radiation were chemical and gravitational. Thomson calculated the age of the Sun for each mechanism and found that gravity gave the largest value, on the order of a few tens of millions of years. Earth could not be older than the Sun, and this was a factor of ten lower than Darwin's estimate of the age of Earth. Using thermodynamics, Kelvin also calculated that the temperature of Earth would have been too high even as recently as a million years ago to allow for life.

Thus, based on the best physics knowledge of Darwin's day, evolution by natural selection was highly suspect. Darwin admitted as much in a letter to his cofounder Wallace: "Thomson's views on the recent age of the world have been for some time one of my sorest troubles." If Thomson's conclusions had been correct, evolution by natural selection would have been falsified.

But Thomson's conclusions were wrong, and Darwin's theory was not falsified. Thomson cannot be faulted, for he used the best information available at the time. However, with the discovery of nuclear energy early in the twentieth century, a new source of energy became known that was far more efficient than either gravity or chemical reactions. This provided a third and more accurate basis for calculating the Sun's age. Furthermore, the natural nuclear radioactivity of Earth provides significant heat and upsets Thomson's calculation for the rate of cooling of Earth.

By mid–twentieth century, the nuclear processes that fuel the Sun were well established and described by theory. By the end of the century, the observation of neutrinos from the Sun (including an observation in which I participated) had directly confirmed the validity of a nuclear source of energy for the Sun and a potential lifetime on the order of ten billion years.[12] Prior to this, radioactive dating also verified that Earth is several billions of years old, and paleontologists have found signs of life going back almost that far.

In short, evolution is as close to being a scientific fact as is possible for any theory given that science is open-ended and no one can predict with certainty what may change in the future. The prospect that evolution by natural selection, at least as a broad mechanism, will be overthrown in the future is about as likely as the prospect of finding out some day that Earth is really flat. Unfortunately, those who regard these scientific facts as a threat to faith have chosen to distort and misrepresent them to the public.

THE CREATIONISTS

By the end of the Victorian era, Darwinian evolution had become widely accepted by the intellectual elite in Britain, including many churchmen as well as scientists. Being mainly confined to scholarly circles, however, this knowledge did not seep substantially into public consciousness. In particular, the great majority of Americans continued to believe in special creation as described in the Bible. They were not hearing from the pulpit what was being discussed in the top divinity schools. Gradually, the press began to take note of the stark contradiction between evolution and common belief, and by 1890 or so the creation/evolution debate had moved into the public arena where it has remained, at or near the surface, to the present day.

As mentioned, what little smoke rises from the pages of biological journals is produced by conflicts over details, not the basic validity of natural selection. For a century now, the creation/evolution war has been fought not in academia but on political and legal fronts, in the media, legislatures, school boards, and courtrooms.

These venues are not particularly noteworthy for their dedication to the establishment of truth. Even with their elaborate formalities, the courts are primarily configured to settle disputes and declare winners and losers. Truth is often sacrificed in the name of another noble ideal—justice.

In the political arena, truth is even lower on the agenda. One can hardly imagine a politician these days gaining reelection by always telling the truth, though they all profess a commitment to "honesty" and "integrity." Similarly, the journalistic media pay lip service to truth, but commercial interests and political correctness rule supreme in that domain.

While truth may be difficult to define, and even more difficult to establish, at least science sets it above all else. The self-correcting nature of the system makes it virtually impossible for scientific fraud to succeed for very long. Scientists are not more or less moral than anyone else, but they are engaged in an enterprise that has evolved methods that make dishonesty a poor strategy for success. Most likely, the great power and success of science is as much, or more, the result of its institutions as of the merit of its individual practitioners.

Once the dangerous idea of evolution moved out of the ivory tower into the open, powerful forces went to work to suppress it. These forces apparently feared an undermining of the social fabric by which they maintained their power. Religion has always been a tool to hammer the masses into line, wielded by those in power to justify their positions by divine right. Asserting their power, in the 1920s the legislatures of the states of Oklahoma, Tennessee, Mississippi, and Arkansas banished the teaching of evolution in public schools.

The tension between opposing forces reached a peak in the famous Scopes "monkey trial" that took place in Dayton, Tennessee, in 1925. A high school teacher named John Thomas Scopes was tried for teaching evolution. His chief prosecutor was renowned orator William Jennings Bryan, three-time losing Democratic candidate for president of the United States. Scopes was defended by celebrated lawyer and freethinker Clarence Darrow.

Bryan argued, "democratically," that a few thousand scientists should not dictate to forty million Christians what should be taught in schools. While he relied on the support of the majority, he also sought scientific backing for his position. A handful of scientists of the time had written antievolutionary tracts, but they proved to be either unwilling or unsuitable witnesses. Bryan could not find two who agreed with each other.[13]

Although Scopes was found guilty and fined $100 (later overturned on a technicality), Darrow, aided by the acerbic pen of reporter H. L. Mencken, triumphed in this public-opinion skirmish. Bryan died a few days after the completion of the trial.

THE CREATION SCIENTISTS

Although creationists continued to work, with some success, to limit the amount of evolution taught in schools, the conflict did not break out into the headlines again until the 1960s, with the rise of *creation science*.

The leader of the new movement was a hydraulic engineer, Henry M. Morris. In 1961 Morris and John C. Whitcomb Jr. published *The Genesis Flood*, which argued for the recent creation of the universe and a worldwide flood that laid down all the geological strata in one year.[14] Although it wildly disagreed with conventional geology, *The Genesis Flood* appeared to nonexperts as a legitimate scientific publication that brought intellectual respectability to the biblical accounts.

This set the pattern for the creationist strategy that has continued to this day: Conduct research and publish articles and books with the goal of establishing an ostensible scientific basis, in the public mind, for the creation story described in Genesis. In 1963 Morris and others formed the Creation Research Society (CRS). Members were required to be Christians and sign a statement of belief accepting the inerrancy of the Bible. Notably, this is incompatible with the unspoken oath of every scientist to pursue the truth whatever it may be, whether one likes the outcome or not. The CRS projects included "scientific expeditions" to search for Noah's ark. In addition, "theoretical studies" were launched to demonstrate the recent origin of Earth, in which the conclusion was known before the data were gathered. A journal, *Creation Research*, was established with the intention of allowing creation scientists to claim publication in a "peer-reviewed" scientific publication. Their peers were, of course, other creation scientists.

In 1972 an Institute for Creation Research (ICR) was established in San Diego under Morris's leadership. He made the purposes of this institution very clear:

> The approach we try to take here [ICR] is to assume that the word of God is the word of God and that God is able to say what He means and means what He says, and that's in the Bible and that is our basis. And then we interpret the scientific data within that framework.[15]

On the current (as of this writing) Web site of the CRS, I found the following in its statement of belief:

> The Bible is the written Word of God, and because it is inspired throughout, all its assertions are historically and scientifically true in the original autographs. To the student of nature this means that the account of origins in Genesis is a factual presentation of simple historical truths.[16]

The record shows that nothing on creation science of scholarly merit has been published in the scientific literature by anyone associated with these organizations. A search by Eugenie Scott and Henry Cole of sixty-eight journals to which scientific creationists could submit articles failed to find a single published paper by people associated with ICR.[17] Of the 135,000 total submissions to these journals from 1980 to 1983, only 18 dealt with empirical or theoretical support for creationism. At the time of the Scott and Cole study, 3 were still pending and 12 had been rejected for poor scholarship. The editors commented that the articles seemed to be written by laymen rather than professional scientists. The situation has not materially changed in the time since this study.

Nevertheless, the creationists established, in their own minds and the minds of many laypeople, the legitimacy of creation science. Sniffing votes in all this, the politicians went back to work exploiting the situation. The new strategy was no longer to eliminate the teaching of evolution, an approach that the U.S. Supreme Court had declared unconstitutional in 1968. Rather, equal time would be demanded for what creationists saw as two alternative scientific models, *creation* and *evolution*. The line became that it is dogmatic and thus very unscientific to teach a single model of human origins—evolution—when that model is "deeply flawed" and creation science is a viable alternative.

The use of the term "model" here, in place of "theory," needs elaboration. While evolution is often excoriated in public debates as "theory and not fact," sophisticated creation scientists do not make the common error of equating theory with "speculation." They understand that the label *scientific theory* is applied only to a well-established body of knowledge that meets certain stringent criteria, although, as we will see, no clear consensus exists among philosophers of science on what precisely these criteria should be. In this regard, creation scientists argue that neither creation nor evolution is a legitimate scientific theory. Rather, they are simply models that one can chose between based on the evidence. As Duane T. Gish presents this view:

> Thus, for a theory to qualify as a scientific theory, it must be supported by events or processes that can be observed to occur, and the theory must be useful in predicting the outcome of future natural phenomena or laboratory experiments. An additional limitation usually imposed is that the theory be capable of falsification; that is, one must be able to conceive some experiment the failure of which would disprove the theory. It is on the basis of such criteria that most evolutionists insist that creation be refused consideration as a possible explanation for origins. Creation has not been witnessed by human observers, it cannot be tested scientifically, and as a theory it is nonfalsifiable.
>
> The general theory of evolution (molecules-to-man theory) also fails to meet all three of these criteria.[18]

Creationists contend that evolution can never be established since it was not witnessed by human observers. But neither have the early big bang or live dinosaurs been observed. Humans cannot observe electrons inside atoms or quarks inside atomic nuclei. Yet these are all phenomena that are (1) supported by events or processes that have been observed to occur, (2) useful in predicting the outcome of future natural phenomena or laboratory experiments, and (3) described by falsifiable theories. Certainly, evolution also meets Gish's criteria. As we saw above, it was even falsifiable at the time Darwin and Wallace first made their proposal of the natural selection mechanism.

The use of the term "model" is very common in science. In physics, for example, we talk about the *standard model* of elementary particles and forces. While that was an appropriate designation when it was first developed in the 1970s, the standard model by today has been so successful in describing observations that it certainly merits being formally recognized as a legitimate *theory*. Similarly, the big bang model in cosmology is sufficiently well established that it can be called the *big bang theory*. This does not imply that these theories will never be supplanted or refuted, just that they fit all the data we now possess with a high degree of reliability. So a "model" represents a step on the way to becoming a "theory." It may fail to make that final step, but even so, it remains a legitimate part of the scientific process.

By calling their proposal a model rather than a theory, the creation scientists were able to evade the application of too-strict criteria and still claim that what they were doing was science. In order to avoid constitutional problems involving church-state separation, Morris urged that public schools teach only the "scientific aspects of creationism." In 1974 the ICR produced a textbook, *Scientific Creationism*, that had one edition for public schools and another for Christian schools, which contained an extra chapter on "Creation According to Scriptures."[19] As we will see with current efforts, the sectarian motives of the creationists have never been heavily veiled.

Besides Morris, the other big gun of the creation science movement of the same period was Duane Gish. Holding a doctorate in biochemistry, Gish is a master debater. In the 1980s he traveled coast-to-coast confronting, head-to-head, biologists who were often less skilled in the tactics of public disputation. These debates were usually conducted before audiences of hundreds, even thousands, recruited from church congregations, who in large majority supported Gish's position with loud cheers. The debate format allowed little opportunity for discussion of the complex scientific issues involved. Though often conducted on college campuses to give them an aura of academic respectability, these debates bore no resemblance to the type of collegial discussion that characterizes normal scientific discourse. While scientific disputes can become quite heated, they never degenerate into popularity contests that are settled by the loudest applause, as was the case in these debates.

The popular success of creation science under Morris's and Gish's leadership soon resulted in further legislative action. In the early 1980s, Arkansas, Louisiana, and various local school boards adopted the "dual model" approach and passed laws mandating equal treatment. However, proevolution forces quickly mobilized. The American Civil Liberties Union (ACLU) filed a federal suit against the state of Arkansas asking that Act 590, the "Balanced Treatment for Creation-Science and Evolution-Science Act," be declared unconstitutional. The ACLU provided attorneys for an assortment of plaintiffs that notably included bishops and clergy from a wide range of religious groups—Catholic, Protestant, and Jewish—as well as teachers and parents.

The trial was held in Little Rock in December 1981, Judge William R. Overton presiding without a jury. In pressing its side of the argument, the ACLU was able to gather an impressive list of expert witnesses, including famed paleontologist Stephen Jay Gould, geneticist Francisco Ayala, and philosopher Michael Ruse. The state chose not to call Morris and Gish as witnesses and did not put on a spirited defense.[20] The creationists were routed. Gould said that they had a victory celebration two days into the two week trial.

On January 5, 1982, Judge Overton ruled against the State of Arkansas, tossing out Act 590. He based his decision on a number of precedents, including the 1971 case *Lemon* v. *Kurtzman* that produced the famous threefold "Lemon test" for determining whether a law meets the Establishment Clause of the U.S. Constitution:

First, the statute must have a secular legislative purpose; second, its principle or primary effect must be one that neither advances nor inhibits religion . . . ; finally, the statute must not foster "an excessive government entanglement with religion."[21]

Overton declared that the evidence strongly confirmed the sectarian purpose of Act 590.[22] He stated that the statute amounted to "a religious crusade, coupled with a desire to conceal this fact." He found, "the evidence is overwhelming that both the purpose and effect of Act 590 is the advancement of religion in the public schools."

Besides resolving this constitutional issue, which was sufficient to sink Act 590, Judge Overton made some additional rulings concerning whether "creation science" is really science and what constitutes science. He observed that the methodology employed by the creationists

> is indicative that their work is not science. A scientific theory must be tentative and always subject to revision or abandonment in light of the facts that are inconsistent with, or falsify, the theory. A theory that is by its own terms dogmatic, absolutist, and never subject to revision is not a scientific theory.[23]

To demonstrate that this was not what creation scientists practice, Overton noted that they "do not take data [and] weigh it against the opposing scientific data" to reach their conclusions. He quotes Morris: "If man wishes to know anything about Creation (the time of Creation, the duration of Creation, the order of Creation, the methods of Creation, or anything else) his sole source of true information is that of divine revelation."[24]

The State of Arkansas decided not to appeal. A similar "equal time" Louisiana law was struck down by the U.S. Supreme Court in 1987 on the grounds that it promoted religion by advancing the view that a supernatural being created the universe (*Edwards* v. *Aguillard*).[25] The Court also ruled that science education would be compromised if schools were forbidden to teach evolution.

Antievolution politicians did not give up, however. The recent strategy has been to enact laws requiring textbook disclaimers.

One such disclaimer prepared by the Louisiana Tangipahoa Board of Education in 1994 read:

> Whenever, in classes of elementary or high school, the scientific theory of evolution is to be presented, whether from textbook, workbook, pamphlet, or other written material, or oral presentation, the following statement shall be quoted immediately before the unit of study begins as a disclaimer from endorsement of such theory.
>
> It is hereby recognized by the Tangipahoa Board of Education, that the lesson to be presented, regarding the origin of life and matter, is known as the Scientific Theory of Evolution and should be presented to inform students of the scientific concept and not intended to influence or dissuade the biblical version of Creation or any other concept.

With the ACLU again coordinating the evolutionists' legal response, in 1997 the disclaimer was ruled unconstitutional in state court. On June 19, 2000, the U.S. Supreme Court voted six-to-three not to hear an appeal of this ruling. By mentioning the "biblical version of Creation," the disclaimer failed the Lemon test. However, Justice Antonin Scalia's dissent, concurred to by Chief Justice William Rehnquist and Justice Clarence Thomas, argued that the biblical reference "is only an illustrative example."[26] This has left room for further attempts at disclaimers, so the story is far from over.[27]

BUT IS IT SCIENCE?

After the 1983 Arkansas case, a significant dispute on the validity of Judge Overton's decision arose among philosophers of science. These philosophers agreed that creation science should not be taught in schools—but for different reasons.

As mentioned, the judge ruled that creation science was, in fact, not science. In this he relied heavily on the testimony of prosecution expert witness, philosopher Michael Ruse. Ruse had written on the philosophical implications of evolution and so was an appropriate choice to take the stand in Arkansas. However, another well-known philosopher of science, and former colleague of mine at the University of Hawaii, Larry Laudan, disagreed strongly with Overton and Ruse.[28] Laudan has spent years studying the so-called demarcation problem, which occupied philosophers of science for a good part of the twentieth century.[29] This problem arises from the attempt by philosophers to agree on a set of criteria that clearly distinguishes science from nonscience.

While most practicing scientists think they have a good idea about what differentiates science from nonscience, neither philosophers nor scientists have ever been able to cast this into a formal principle that can be applied to all cases. Laudan and most other philosophers would agree that the criteria used by Gish described above are inadequate. Following Ruse's counsel, Judge Overton arrived at the following five criteria for defining whether or not something is scientific:[30]

(1) it is guided by natural law;
(2) it has to be explained by reference to natural law;
(3) it is testable against the empirical world;
(4) its conclusions are tentative, that is, are not necessarily the final word; and
(5) it is falsifiable.

Laudan demurred: "At various key points in the opinion, creationism is charged with being untestable, dogmatic (and thus non-tentative), and unfalsifiable. All three charges are of dubious merit."[31] He pointed out that creationism makes a wide range of assertions that can be tested empirically, such as that Earth is of a recent origin (6,000–20,000 years) and that Earth's geological features are the result of the Noachian deluge. They make other factual claims based on the Bible, such as the cocreation of animals and humans. These are all testable and, in fact, have failed the tests. Furthermore, creation scientists have modified their positions over time.

Let me try to indicate the problems with Overton's criteria, which are characteristic of most other attempts to define science. One would like criteria that are both *necessary* and *sufficient*, so that any statement could be tested against these criteria and declared scientific or not.

Overton's first two criteria, listed above, involve natural law. While it is true that up until now science has been confined to natural phenomena, as opposed to what would be labeled supernatural or simply nonnatural, nothing demands that this always be the case. If scientific instruments uncover nonnatural phenomena, scientists would study them just as they currently study natural phenomena. All they need are the data.

In any case, even if dealing only with natural phenomena were a necessary condition for an activity to be scientific, it is not a sufficient condition. Many commonly accepted nonscientific activities, such as plumbing and basket-weaving, deal with natural phenomena. Furthermore, plumbing, at least, is certainly guided by natural laws such as Bernoulli's principle (the pressure in a stream of fluid is reduced as the speed of the flow is increased).

What about empirical testability and falsifiability? No doubt science deals with the empirical, that is, observational or experimental data, and this may be regarded as a necessary condition for being labeled science. But this, too, is not sufficient, unless you want to call astrology, palm reading, and every other occult practice scientific. Astrology, for example, makes falsifiable predictions that have been falsified. The falsification criterion would say that astrology is science, just wrong science. While that might be acceptable to some, most philosophers of science think this opens up so much room for activities to be called science that the designation would cease to carry much meaning.

Laudan accused Overton of applying a false dichotomy (we will encounter many in this book): "Since Creationism is not 'science,' it must be religion."[32] He predicted that the Arkansas decision would come back to haunt science by "perpetuating and canonizing a false stereotype on what science is and how it works." The decision left plenty of space for creationists to rearrange their arguments to meet Judge Overton's now precedent-making legal criteria and, as we will see below, they have exploited this possibility. Rather than rely on questionable criteria to declare the whole of creationism nonscientific, Laudan argued that we should examine its claims one-by-one, see which can be fairly tested against the evidence, and then proceed to make those tests.

Reacting to Laudan's criticism, Ruse defended Overton's ruling by pointing out that the plaintiffs needed to show that creation science was religion in order to have Act 590 ruled invalid on constitutional grounds.[33] There was ample evidence from the creation scientists' own writings that a religious rather than secular purpose would be served by teaching their version of creationism, which is clearly based on the Bible, in public school science classes. He pointed out that the Constitution does not bar the teaching of weak science, so the plaintiff's tactic was to show that it was not just bad science but not science at all.

Laudan responded by reiterating that the emphasis on defining science did more harm than good.[34] In this, he was supported by another philosopher, Philip Quinn, who analyzed the case and concluded that "scientists and their friends should derive little comfort from the outcome."[35]

INTELLIGENT DESIGN:
THE NEW STEALTH CREATIONISM

Creationists responded quickly to the legal developments in Arkansas and a new version of creation science soon emerged into the spotlight. This re-creation of creation science parades under a banner labeled *intelligent design*. Just as advocates of the Morris and Gish brand of creation science learned from the mistakes of those who sought to outlaw the teaching of evolution outright, the new intelligent design creationists learned from the mistakes of Morris and Gish. While intelligent design differs in substantial ways from its previous incarnations, unabashed religious creationism it remains.

The intelligent design creationists learned three lessons from the history of their movement. First, you must not appear to be promoting any one particular sectarian belief system, but simply presenting "evidence" for a generic creator that need not even be supernatural. Second, do a better job than previous creation scientists in avoiding claims that are outrageous or easily falsified, such as Earth being only 6,000 years old. Third, and here is where Laudan's prediction has come true, argue that conventional science has a built-in dogmatic attachment to naturalism which prevents it from even considering supernatural causes. Thus, it is "censorship" to prevent intelligent design from being taught in public school science classes.

Despite these adjustments in strategy, intelligent design proponents are almost exclusively Christians and have not tried to hide the identity of the creator they have in mind. Their leading theorist is mathematician/theologian William Dembski. While producing various mathematical formulas for the scientific detection of intelligent design in nature, he writes that "any view of the sciences that leaves Christ out of the picture must be seen as fundamentally deficient."[36] Furthermore, Dembski and his colleagues have only partially succeeded in avoiding the appearance of doing bad science. While they have gone one better than previous creation scientists in persuading many nonscientists, academics as well as laypeople, of the merit of their point of view, they have had no success in convincing the experts in the fields they address. In over a decade since the notion of intelligent design was introduced, not a single paper on the subject has been published in a mainstream scientific journal. In his regular column in the *Boston Globe*, Chet Raymo noted:

> The new advocates of ID [intelligent design] ask that their ideas be judged by scientific, not religious, criteria. OK, let's see how well ID stacks up as a scientific alternative to Darwinism. To gauge how well ID is doing as a platform for scientific research, I logged into the best database of the biological literature. A search for keyword "evolution" yielded 24,000 hits in the last decade. A search for "intelligent design" yielded not a single piece of research. Evolution by natural selection remains the basis of every successful biological research program.[37]

As we will see, intelligent design has implications that go beyond the creation/evolution controversy. For the rest of this chapter, however, I want to continue to discuss the events that have continued to unfold in the political and legal arenas. Unsurprisingly, the current leading spokesperson of the new creationists is a lawyer.

PROSECUTING NATURALISM

At the time of this writing, the battle against evolution—or more generally, "naturalism"—is being spearheaded by retired University of California at Berkeley criminal law professor Phillip Johnson. He has a no-holds-barred, winner-take-all approach that one would expect in the prosecution of a mass murderer or a child molester.

Along with Dembski and most of the other leaders in the intelligent design movement, Johnson is associated with the Center for Science and Culture, which is part of the Discovery Institute mentioned in the preface. They have developed what is called call the "wedge strategy" with the goal of cutting off materialism at its source, namely scientific naturalism.[38] The approach of the wedge strategy has been developed in several books by Johnson, most recently *The Wedge of Truth: Splitting the Foundations of Naturalism*.[39]

Johnson expresses no sympathy for those theists, such as the leaders of the Catholic church and many liberal Protestant theologians, who view evolution as fully compatible with belief in God:

> [Evolution] doesn't mean God-guided, gradual creation. It means unguided, purposeless change. The Darwinian theory doesn't say that God created slowly. It says that naturalistic evolution is the creator, and so God had nothing to do with it.[40]

Johnson sees evolutionary naturalism as a cause of many of the "evils" of modern society including homosexuality, abortion, pornography, divorce, and genocide, as if the world had none of these before Darwin came along.[41]

Johnson's description of evolution in his writing has been characterized by at least one biologist as a "crude caricature."[42] But, clearly, biologists are not his audience. As philosopher Robert T. Pennock described Johnson's method:

> He knows how to draw upon his strengths and makes a classic courtroom move of shifting the locus of argument in a way that seeks to undermine the expert testimony of his scientist adversaries. His key argument is broadly philosophical, but Johnson also uses his considerable rhetorical skills to try to turn the tables on scientists by portraying them as naïvely doctrinaire and intolerant, while portraying creationists as rational and fair-minded skeptics.[43]

This evaluation is echoed by biologist, and theist, Kenneth Miller:

> When I first read Phillip Johnson's book, *Darwin on Trial*, I read it as a scientist and it puzzled me. In every chapter he attacked what he considered to be a weak spot in evolutionary theory, implying in each and every case that there might be another explanation, a better one than evolution. This is a common strategy in a scientific argument. As I neared the end of the book, I expected Johnson to do what one of my scientific colleagues would do at the conclusion of a provocative seminar—to lift the curtain and reveal that better explanation. Like any scientist, I expected him to present a model that would fit the data more precisely, a model that would possess powers of explanation and prediction well beyond the theory he had attacked. But Johnson did nothing of the kind.

Miller continues:

> Gradually I realized that the case he and his associates bring against evolution is *not* a scientific case at all but a legal brief. The goal of his brief is to raise reasonable doubt, to create a climate in which the intellectual claims of evolution seem shaky, even unreasonable. What it never does is present an alternative—any alternative—to the seamless integration of theory and natural history provided by evolution.[44]

Furthermore, Johnson has not come up with a scientifically novel case against evolution. Instead, he simply recites the usual arguments one hears over and over again in creationist literature despite having been refuted time and time again. His central argument seems to be that "in our universal experience unintelligent material processes do not create life."[45] I would daresay that human experience is hardly "universal," living as we do in a tiny region of space and time. Human experience tells us that the world is flat.

Unlike those theists who at least pay lip service to science and scientific method, Johnson is out to convict science of fraud in the court of public opinion:

> Scientific naturalism is a story that reduces reality to physical particles and impersonal laws, [and] portrays life as a meaningless competition among organisms that exist only to survive and reproduce.[46]

Pennock summarizes Johnson's case against science as follows:

> Evolution is a naturalistic theory that denies by fiat any supernatural intervention. The scientific evidence for evolution is weak, but the philosophical assumption of naturalism dogmatically disallows consideration of the creationist's alternative explanation of the biological world. Therefore, if divine interventions were not ruled out of court, creationism would win over evolution.[47]

Johnson equates naturalism with materialism: "*Naturalism* and *materialism* mean essentially the same thing."[48] Thus a creationist, Johnson claims, has no chance to present his views in an academic environment:

> In our greatest universities, naturalism—the doctrine that nature is "all there is"—is the virtually unquestioned assumption that underlies not only natural science but intellectual work of all kinds.[49]

Pennock points out that Johnson is presenting the argument as a dichotomy in which creationism and evolution are the only alternatives. As we saw earlier, this is a common creationist tactic that does not work. If one wishes to establish divine creation, it does not suffice for one to simply refute evolution. One must argue independently for the validity of creationism. Creationism is not declared victorious by default.

Johnson does not ignore this issue. But he does not view science as the only method by which the validity of creation can be established. He mentions sacred books and mystical experiences—that is, divine revelation—as other sources of evidence, what we have called *mythos*. He also adopts the ancient argument from design: "From the very fact the universe is on the whole orderly, in a manner comprehensible to our intellect, is evidence that we and it were fashioned by a common intelligence."[50]

In Johnson's recent book, *The Wedge of Truth*, he finally provides an answer for the question raised by Miller above and often asked of Johnson after his speeches: If the blind process of natural selection was not the mechanism for the development of life, then what was the mechanism? Johnson tells us that the mechanism was the *logos* of the Gospel of John. Here he is clearly not using the term to mean the rational, scientific route to knowledge, as I have used it, but some active, supernatural force in the cosmos—the Word.

In a review of *The Wedge of Truth*, Jesuit theologian Edward T. Oakes notes that Johnson leaves aside "the uncomfortable fact that no biblical or doxological text in either Judaism or Christianity praises God as the Celestial Cell Constructor or Divine *Bauplan* [German for "blueprint"] Architect."

Oakes adds:

Throughout Johnson's book, and indeed throughout all his writings on this subject, there lurks, like the Ghost of Christmas Past, clanking chains and all, the unexorcised spirit of the Anglican Archdeacon William Paley (1743–1805), whose lucubrations on the "clockmaker God" so impressed Darwin in his undergraduate days. In my opinion, anyone who follows that hyper-cheerful, almost Candide-like clergyman down the designer road is asking for trouble later on; and indeed once Darwin became a naturalist (in the nineteenth-century meaning of that word: an investigator and collector of species), his departure from Christian orthodoxy was well-nigh inevitable.[51]

MEETING THE CHALLENGE

It is not my intention in this book to review all the arguments for and against evolution. That takes a book in itself, and many such books are already available, along with numerous other resources such as journal articles and Web sites.[52] My purpose in this chapter is to show that many Christians regard evolution as such a great threat to their faith that it must be fought against by every means at their disposal, even when those means are less than scrupulous. While antievolution fanaticism does not animate the majority, and we will later be meeting a number of Christian theologians and scientists who have woven evolution into their belief systems, the antievolution crusade in the United States is too powerful to ignore. The effects can easily spill over into a general assault on science to its great detriment and to the great detriment of civilization.

Antievolutionism cannot be dismissed simply because of its lack of scientific merit coupled with the often ridiculous and downright false assertions made by its promoters. These shortcomings do not translate into impotence when the populace at large lacks the sophistication to recognize that they are being manipulated. The emotionally charged creation crusade remains a robust movement in the United States. Polls indicate that most Americans think that evolution should be taught in schools, but they also misconstrue creation theory as a viable scientific alternative.[53] This is what they hear from many pulpits and read in many books. So many are willing to allow creation science a place in the science curriculum. With both public support and money, they can easily attract politicians only too willing to make public policies that support creationist views in return for campaign contributions and votes. If scientists want to avoid being stigmatized as

dogmatically atheistic, it is imperative that they respond in a carefully controlled fashion which is consistent with their avowed
claim of seeking the truth—whatever it may be.

When Phillip Johnson accuses science of refusing to consider
nonnatural causes, a common response from scientists has been to
argue that science deals only with the natural. This is a dogmatic
response that plays right into Johnson's hands. Recall, from the
discussion of the Arkansas trial, that the first two of Judge
Overton's criteria for science demanded that it be guided by and
explained with reference to natural law. Even if that has been the
case until now, it need not be so forever. Scientists should accept
Johnson's challenge and reply that they are open to the consideration of nonnatural causes. As Pennock and others have pointed
out, the naturalism of science is *methodological* and not necessarily
ontological.[54] If the evidence for nonnatural causes is there, as many
theists are now claiming, then scientists will proceed to examine
that evidence fairly and openly, that is, by submitting it to the glare
of scientific method.

While naturalism has proven to be a useful working assumption in science, and economy of thought demands that all natural
causes be exhausted before adding new assumptions, scientists
have no emotional or dogmatic attachment to this assumption.
Show them the data. Most would be thrilled if a whole new world
opened up to explore by empirical means. Unfortunately, all
Johnson and his colleagues have done so far is attack evolution,
leaving a gaping wound of ignorance with only the diaphanous
band aids of mysticism to cover them. They have not presented
any alternative mechanism that can be tested against the data in
normal scientific fashion.

If theists say they want to play on the science ball field, how
can scientists object? This is the scientists' game, and they know
the rules very well. Furthermore, it gives scientists the home-field
advantage. If a theist makes an empirical claim, then scientists can
investigate that claim scientifically. At least that will demonstrate
their willingness to objectively search for the truth without preconceptions.

After seeing some of the tactics described in this chapter, however, scientists ought to be suspiciously alert and vigorously object whenever creationists attempt to move the discussion to another venue. No matter how smart scientists think they are, creationists have a good chance of winning any contest when the stadium is packed with their cheering fans and the umpires do not enforce the rules. Only by sticking strictly to the science can we expect a rational outcome.

In 1979 Duane Gish wrote:

> Whether evolution happened or not can only be decided, scientifically, established by the discovery of fossilized remains of representative samples of those intermediate types which have been postulated on the basis of indirect evidence. . . . As a matter of fact, the discovery of five or six of the transitional forms scattered through time would be sufficient to document evolution.[55]

Thus, Gish presents evolutionists with a demand to provide specific empirical evidence that if met, he promises, will provide sufficiently compelling evidence to convince him and his fellow creationists of the validity of evolution.

Gish's challenge has been taken up by many biologists and paleontologists.[56] Unsurprisingly, the vast fund of data they have provided has not so far converted Gish or any other creationists to evolutionism.

The alleged absence of "transitional forms" has been a longstanding argument made by creationists against evolution. In the evolutionary scheme, every species is in some sense a transitional form between two other species. Indeed, *Homo sapiens* is a transitional form. We are not likely to remain unchanged in the next million years. Paleontologists have presented numerous examples of fossils which show indications that the organisms lie between other species. The most famous example is *Archaeopteryx*, which exhibits features of both bird and reptile. However, creationists argue that it is a bird by definition and so not a transitional form. Others call it a reptile.[57]

Geologist, and theist, Keith Miller has given this nice summary of the situation with respect to whales:

One of the currently best documents of transitional fossil sequences is for the evolution of whales from a terrestrial hoofed animal. There are now twenty-six fossil species of primitive whales known that have been assigned to four families. These together provide an impressive fossil sequence of transitional forms—the "walking whales." This progression of fossil forms shows a clear trend from more terrestrial to fully marine adaptations. Associated with this is a progressive change in swimming from limb propulsion to tail undulation. The fossil record of early whales is even more impressive because of its paleoenvironmental and geographic aspects. Not only do the fossils occur in the correct chronological order, but they are found in progressively more marine settings.[58]

Raymond Sutera has provided a comprehensive survey of the overwhelming evidence that whales evolved from terrestrial mammals.[59] The evidence comes from eight independent aspects of scientific study: paleontological, morphological, molecular biological, vestigial, embryological, geochemical, paleoenvironmental, and chronological.

Numerous other examples can be found in the record, but obviously they can always be defined away by specious arguments. Died-in-the-wool creationists will never be convinced by any scientific data, despite Gish's promise. They already know the truth, and the data must be made to conform to it. But this does not mean that their assertions should be ignored. Scientists must continue to respond to make sure that the public record is not limited to one side of the dispute.

When all is said and done, however, the creation/evolution debate has little to do with the debate over scientific evidence for the existence or nonexistence of God. As I have noted, it is a false dichotomy to think that by debunking evolution one is providing evidence for a creator with the humanlike features of the Judeo-Christian-Islamic God. If this God exists, his face should be discernable on Earth and in the Heavens. Only the childish self-centeredness of humanity leads us to seek evidence for purpose in the thin layer of carbon that coats the surface of a minor planet.

NOTES

1. Thomas Kuhn, *The Structure of Scientific Revolutions* (Chicago: University of Chicago Press, 1970).

2. Steven Weinberg, "The Revolution That Didn't Happen," *New York Review of Books*, October 8, 1998.

3. Quotations taken from the Revised Standard Version.

4. Daniel C. Dennett, *Darwin's Dangerous Idea: Evolution and the Meanings of Life* (New York: Simon & Schuster, 1995).

5. Andrew Dickson White, *A History of the Warfare of Science with Theology in Christendom* (1896; reprint, Amherst, N.Y.: Prometheus Books, 1993), pp. 70–88.

6. Cyril Bibby, *T. H. Huxley: Scientist, Humanist, and Educator* (London: Watts, 1959), p. 259. Note that the exact nature of their exchange is still debated.

7. Pius XII, *Humani Generis*, August 12, 1950.

8. John Paul II, address to the Academy of Sciences, October 28, 1986, *L'Osservatore Romano* English ed., November 24, 1986, p. 22. The English translation is currently online at www.cin.org/users/james/files/message.htm.

9. Andrew Petto, "Who Is Fooling Pope John Paul II?" *Reviews of the National Center for Science Education* 18, no. 4 (1998): 23–24.

10. John Paul II, address to the Academy of Sciences.

11. Th. Dobzhansky and O. Pavlovsky, "An Experimentally Created Incipient Species of Drosophila," *Nature* 23 (1971): 289–92; T. Mosquin, "Evidence for Autopolyploidy in *Epilobium angustifolium* (Onaagraceae)," *Evolution* 21 (1967): 713–19; S. Stanley, *Macroevolution: Pattern and Process* (San Francisco: W. H. Freeman and Company, 1979), p. 41; E. Mayr, *Populations, Species, and Evolution* (Cambridge: Harvard University Press, 1970), p. 348.

12. John N. Bahcall, "How the Sun Shines," Nobel e-Museum, The Nobel Foundation [online], www.nobel.se/physics/articles [June 19, 2000].

13. Ronald L. Numbers, *The Creationists* (New York: Alfred A. Knopf, 1992), pp. 72–73.

14. John C. Whitcomb Jr. and Henry M. Morris, *The Genesis Flood: The Biblical Records and Its Scientific Implications* (Philadelphia: Presbyterian and Reformed Publishing Co, 1961).

15. Morris as quoted in Brian J. Alters, "A Content Analysis of the Institute for Creation Research's Institute on Scientific Creationism," *Cre-*

ation/Evolution 15, no. 2 (1995): 1–15. Also see this reference for more details on the ICR.

16. Creation Research Society [online], www.creationresearch.org.

17. Eugenie C. Scott and Henry P. Cole, "The Elusive Basis of Creation 'Science,'" *Quarterly Review of Biology* 60, no. 1 (1985): 21–30.

18. Duane T. Gish, "Creation, Evolution, and the Historical Evidence," *American Biology Teacher* (March 1973): 132–40.

19. Henry M. Morris, ed., *Scientific Creationism* (El Cajon, Calif.: Creation-Life Publishers, 1974).

20. Michael Ruse, ed., *But Is It Science? The Philosophical Question in the Creation/Evolution Controversy* (Amherst, N.Y.: Prometheus Books, 1996).

21. *Stone v. Graham*, 449 US 39 (1980).

22. William R. Overton, United States District Court Opinion, *McLean v. Arkansas*, 1982. Reprinted in Ruse, *But Is It Science?* pp. 307–31.

23. Ibid.

24. Morris, Plaintiffs' Exhibit 312 in *McLean v. Arkansas*.

25. *Edwards v. Aguillard*, 482 U.S. 578 (1987).

26. Antonin Scalia. Dissenting Opinion, 120 U.S. Supreme Court 2706, 2000.

27. For the view from the U.S. National Academy of Sciences, see National Academy of Sciences, *Teaching about Evolution and the Nature of Science* (Washington, D.C.: National Academy Press, 1998).

28. Larry Laudan, "Science at the Bar—Causes for Concern," *Science, Technology, & Human Values* 7, no. 41 (1982): 16–19. Reprinted in Ruse, *But Is It Science*, pp. 351–55 and in Larry Laudan, *Beyond Positivism and Relativism: Theory, Method, and Evidence* (Boulder, Colo. and Oxford: Westview Press, 1996), pp. 223–27.

29. Larry Laudan, "The Demise of the Demarcation Problem," in *Physics, Philosophy, and Psychoanalysis*, ed. R. S. Cohen and L. Laudan (Dortrecht, Holland: Reidel, 1983), pp. 111–27. Reprinted in Ruse, *But Is It Science*, pp. 337–50 and Laudan, *Science and Values: The Aims of Science and Their Role in the Scientific Debate* (Berkeley and Los Angeles: University of California Press, 1984).

30. Overton, *McLean v. Arkansas*.

31. Laudan, "Science at the Bar."

32. Ibid.

33. Ruse, *But Is It Science?*

34. Laudan, "Science at the Bar."

35. Philip L. Quinn, "The Philosopher of Science as an Expert Witness," in *Science and Reality: Recent Work on the Philosophy of Science*, ed. James T. Cushing, C. F. Delaney, and Gary M. Gutting (South Bend, Ind.: University of Notre Dame Press, 1984). Reprinted in Ruse, *But Is It Science?*

36. William Dembski, *Intelligent Design: The Bridge between Science and Theology* (Downers Grove, Ill.: InterVarsity Press, 1999), p. 298.

37. Chet Raymo, "Science Musings: Intelligent Design Does Not Compute," *Boston Globe*, June 19, 2001.

38. [online] http:// www.public.asu.edu/~jmlynch/idt/wedge.html.

39. Phillip E. Johnson, *The Wedge of Truth: Splitting the Foundations of Naturalism* (Downers Grove, Ill.: InterVarsity Press, 2001).

40. Phillip E. Johnson, *Defeating Darwinism by Opening Minds* (Downers Grove, Ill.: InterVarsity Press, 1997), p. 16.

41. Phillip E. Johnson, *Reason in the Balance: The Case against Naturalism in Science, Law, and Education* (Downers Grove, Ill.: InterVarsity Press, 1995). The reference to homosexuality is page 2; divorce, page 41; and genocide, page 144.

42. William B. Provine, "Response to Phillip Johnson," in *Evolution as Dogma: The Establishment of Naturalism*, ed. P. E. Johnson (Dallas: Haughton Publishing Co., 1990).

43. Robert T. Pennock, *Tower of Babel: The Evidence against the New Creationism* (Cambridge: MIT Press, 1999), p. 184.

44. Kenneth R. Miller, *Finding Darwin's God: A Scientist's Search for a Common Ground between God and Evolution* (New York: HarperCollins, 1999), p. 123.

45. Johnson, *Reason in the Balance*, p. 108.

46. Ibid., p. 197.

47. Pennock, *Tower of Babel*, p. 185.

48. Phillip E. Johnson, *Defeating Darwinism by Opening Minds* (Downers Grove, Ill.: InterVarsity Press, 1997), p. 15.

49. Johnson, *Reason in the Balance*, p. 7.

50. Phillip E. Johnson, *Evolution as Dogma: The Establishment of Naturalism* (Dallas: Haughton Publishing Co., 1990), p. 13.

51. Edward T. Oakes, S.J., "Newman, Yes; Paley, No," review of *The Wedge of Truth*, by Phillip E. Johnson, *First Things* 109 (January 2001): 48–52.

52. Rather than make a long list of Web addresses, which would be out of date the moment this book is published, the reader is urged to use

the search engines on the Web to find the most useful sites. These include many links to articles that can be downloaded, FAQs (lists of answers to frequently asked questions), and lists of the latest references.

53. James Glanz, "Both Evolution and Creationism Find Support," *Akron Beacon Journal*, March 11, 2000; Gallup poll February 14, 2001 [online], http://www.gallup.com/poll/releases/pro10214C.asp.

54. Pennock, *Tower of Babel*.

55. Duane T. Gish, *Evolution—The Fossils Say No!* (San Diego: Creation-Life, 1979), p. 49.

56. L. Beverly Halstead, "Evolution—The Fossils Say Yes!" in *Science and Creationism*, ed. A.Montagu (New York and Oxford: Oxford University Press, 1984); Daniel G. Blackburn, "Paleontology Meets the Creationist Challenge," *Creation/Evolution* 36 (1995): 26–38; Miller, *Finding Darwin's God*.

57. For a review of recent developments in the evolution of birds, see Kevin Padian, "Dinosaurs and Birds—an Update," *Reports of the National Center for Science Education* 20, no. 5 (2000): 28–30.

58. Keith Miller, private e-mail communication.

59. Raymond Sutera, "The Origin of Whales and the Power of Independent Evidence," *Reports of the National Center for Science Education* 20, no. 5 (2000): 33–41.

THREE

NO REASON
TO BELIEVE

The ancient covenant is in pieces; man at last knows that he is alone in the universe's unfeeling immensity, out of which he emerged only by chance. His destiny is nowhere spelled out, nor is his duty. The kingdom above or the darkness below: it is for him to choose.

—Jacques Monod, 1970

WHO BELIEVES? AND, WHY?

Polls indicate that 90 percent of Americans believe in a personal God, when defined along traditional Judeo-Christian-Islamic lines.[1] Perhaps another 5 percent believe in some more nebulous superior being or "universal spirit." In his 2000 book, *How We Believe: The Search for God in an Age of Science*, Skeptics Society founder Michael Shermer reports on a survey he conducted, with

Frank Sulloway, on the religious attitudes of a sample of almost one thousand Americans.[2] This poll indicated 64 percent belief, lower than other surveys, which Shermer attributes to a higher-than-average educational level of those who participated in the survey.

Despite the Shermer-Sulloway sample being admittedly non-representative, one result was interesting and perhaps a bit surprising. When believers were asked the main reason why they believed, 28.8 percent said it was because of arguments based on good design, natural beauty, perfection, and the complexity of the world or universe. Of the remainder, 20.6 percent said they believed because of "the experience of God in everyday life" or "a feeling that God is in us." Only 10.3 percent admitted they believed primarily because they found it "comforting, relieving, consoling," and it gave "meaning and purpose to life." On the other hand, when asked why they thought *other* people believed in God, 26.3 percent said it was for reasons of comfort and meaning.

So, it seems that half of the people who believe in God say they do so for what might loosely be called "scientific" reasons—conclusions based on their own personal experiences and observations of the world around them. While they presume that most other believers use God to fulfill emotional needs, many individual believers, at least among the better educated, view themselves as being above such childish feelings. They have faith because of what they see as *evidence* for God based on their analysis of the data available to them.

What about that group of professionals most highly trained in the interpretation of observational evidence? A recent survey of the members of the U.S. National Academy of Sciences (NAS) indicates that only 7 percent believe in a personal god.[3] This is dramatically down from 27.7 percent in 1913 and 15 percent in 1933. Personal disbelief, as distinguished from simple nonbelief, among the NAS group now stands at 72.2 percent and doubt or agnosticism at 20.8 percent. Disbelief among physical scientists stands the highest, at 79 percent.

Other surveys, however, suggest that the religiosity of academic scientists as a whole is not much lower than that of the general public.[4] Apparently, unbelief among the nation's science elite is higher than that of the more general population of scientists teaching in U.S. colleges.

No doubt, believing, nonbelieving, and disbelieving scientists all view their personal choices as the result of a rational contemplation of the evidence. On the other hand, despite their individual convictions, most scientists would probably prefer not to see science get mixed up with religion. Many agree with paleontologist Stephen Jay Gould (now deceased) that science and religion have their own domains, what he has termed "nonoverlapping magisteria."[5] Science is supposedly concerned with describing empirical phenomena, and religion is limited to matters of morality and, perhaps, questions of ultimate meaning. However, as philosopher Owen McCleod points out, Gould basically redefines religion to be what is more properly called "philosophical ethics."[6] For example, Gould explicitly excludes creationism from the category of religion because it tries to "impose a dogmatic and idiosyncratic reading of a text upon a factual issue lying within the magisterium of science."[7] Gould's magisteria do not overlap as long as religion avoids science's turf. The problem is, it does.

In a letter to the Church in 1632, Galileo related the opinion of Cardinal Baronius (1538–1607): "The intention of the Holy Ghost is to teach us how one goes to heaven, not how heaven goes." Still, just as this catchy line did not save the greatest scientist since Aristotle from getting entangled with religion, modern scientists are not going to escape that easily either.

Scientists have practical reasons for wishing that religion and science be kept separate. They can see nothing but trouble in bringing religious questions into science, with the likely prospect that nothing will get settled anyway. Research scientists, especially, prefer to be left alone to do their work. They worry about the erosion of public and political support, with a subsequent decrease in funding, if they venture into the deeply divisive issue of religion—especially when their observations tend to support a highly unpopular, atheistic conclusion.

The difficulty with this cautious, hands-off attitude is that the existence of God, at least the personal God of the Judeo-Christian-Islamic traditions which great numbers of people worldwide practice, has empirical consequences and makes claims that, at least in principle, can be tested scientifically. And, if they can, they will be. As Richard Dawkins has put it:

> More generally it is completely unrealistic to claim, as Gould and many others do, that religion keeps itself away from science's turf, restricting itself to morals and values. A universe with a supernatural presence would be a fundamentally and qualitatively different kind of universe from one without. The difference is, inescapably, a scientific difference. Religions make existence claims, and this means scientific claims.

Dawkins adds:

> There is something dishonestly self-serving in the tactic of claiming that all religious beliefs are outside the domain of science. On the one hand, miracle stories and the promise of life after death are used to impress simple people, win converts, and swell congregations. It is precisely their scientific power that gives these stories their popular appeal. But at the same time it is considered below the belt to subject the same stories to the ordinary rigors of scientific criticism: these are religious matters and therefore outside the domain of science. But you cannot have it both ways. At least, religious theorists and apologists should not be allowed to get away with having it both ways. Unfortunately all too many of us, including nonreligious people, are unaccountably ready to let them.[8]

Despite Gould's attempt to divide up the territory, scientific evidence for God's existence is being claimed today by theists, many of whom carry respectable scientific or philosophical credentials. "He," who is neither a "she" nor an "it," supposedly answers prayers and otherwise dramatically affects the outcome of events. If these consequences are as significant as believers say, then the effects should be detectable in properly controlled experiments. With all the observational power of science today, it makes sense that only if God's role in the current universe were negligible would it escape detection. That being the case, scientists cannot simply stick their heads in the sand and ignore the discussion. If they do, they will be preempted on their own turf by people, like the creationists described in chapter 2, who are committed to an agenda other than a dispassionate scientific search for the truth.

Furthermore, traditional religions dominate much of the thinking of humanity and make claims about the nature of reality that often contradict those implied by science. Most sacred scriptures, even when not taken literally, paint a picture of the universe that is very difficult to reconcile with what has been revealed by science. Again, scientists must be prepared to respond in an objective and knowledgeable manner.

EXAMINING THE EVIDENCE

Nowhere is the discrepancy between scientific and religious views more profound than on the matter of the position and destiny of human life. From cosmology to biology, the divine plan for humanity that forms the underlying theme of most traditional religions is not confirmed by scientific observations. Since Copernicus, observational and theoretical cosmology has moved humankind farther and farther from the physical center of the universe to the point where, today, Earth is viewed as a tiny speck in a vast cosmos—less noteworthy than a single grain of sand in the Sahara Desert.

Likewise, biology has found increasingly convincing evidence that humans evolved in a nondirected fashion from simpler life-forms. Cognitive science has moved strongly toward adopting a purely material basis for all thinking, including so-called mystical experiences.[9] Computer science has provided working models of complex, material systems that self-organize and develop solutions to problems without human guidance, often exhibiting greater creativity than humans.[10]

And here is where a new debate has arisen. Many theists have religious notions so deeply ingrained in their thinking that they cannot accept this interpretation of the scientific message. Almost in desperation, they assert the contrary—that modern science now provides *evidence* for the existence of a God who has a special interest in humanity and divine intentions for its future. They insist that this is not just a matter of faith, but scientifically verified fact.

Philosophers in the twentieth century grappled mightily with the question of the verification of scientific statements. In the end, they concluded what most scientists already understood, that no fact about nature can be known with 100 percent certainty. However, this does not necessarily imply complete uncertainty. A spectrum of probabilities for the validity of scientific statements exists in the range between 0 and 100 percent. Given the highly developed state of science today, many statements can be made that approach very close to 100 percent certainty. For example, physics can predict that if you jump off a ten-story building to the ground, you will very likely be killed. Gravity could weaken for a moment to allow you to float to safety, but don't bet your life on it. Every day we put our trust in the power of science to predict events with high likelihood.

If we accept this, then we must also accept as a possibility that science can demonstrate, with probability approaching certainty, the existence of a God whose reality has significant empirical consequences. And if that is possible, so is the opposite possibility, that science can demonstrate, with probability approaching certainty, the nonexistence of a God who has the far-from-negligible empirical consequences implied by the religions that most people on Earth practice.

CREATOR AND COSMOS

Exemplifying one extreme of the theistic scientist position is physicist and astronomer Hugh Ross. At this writing, Ross serves as the president of Reasons to Believe, a nonprofit corporation that produces Christian materials and a weekly radio show. His several books include *The Fingerprint of God, The Creator and the Cosmos,* and *The Genesis Question.*[11]

True to the name of his organization, Ross attempts to provide Christians with scientific reasons to believe in a personal creator. Listening to the radio program, one gets the impression that his audience is comprised mainly of people who are already committed Christians. He, his cohosts, and guests provide their listeners with answers to questions that may be raised as they go about their mission of bringing others into the fold. You don't need to argue from faith, Ross tells them to say. The God of the Bible is a scientific fact. He insists that observational data verify the existence of the Christian God with about the same level of confidence that data can verify anything we label as scientific fact.

I have characterized Ross as exemplifying an extreme position among theistic scientists. However, he is not so extreme as to promote the scientifically unsound notions of the young-Earth creationists and other antievolutionists who we met in chapter 2. They are so far off the scale that their scientific claims need not be taken seriously. Their distortions and misrepresentations of the scientific facts are not consistent with their self-righteous claims of acting to protect all that is good and moral. Still, as I have emphasized, their political muscle cannot be ignored without consequence.

Ross, by contrast, does not dispute scientific results, even as he interprets them very loosely to suit his purpose. He accepts evolution, in some guided form, as well as the vast and ancient universe pictured in modern cosmology. However, when he reviews the evidence from cosmology, Ross concludes that it "determines that the cause of the universe is functionally equivalent to the God of the Bible, a being beyond the matter, energy, space, and time of the cosmos."[12] In this insistence on the basic scientific accuracy of the Bible, Ross remains in conflict with other science-theists who are trying to reconcile science and religion.

Ross's primary conclusion, that the God of the Bible can be found in science, is based on his interpretations of modern physics and cosmology. According to the current consensus of cosmologists, our universe exploded from a tiny point in space some 13 billion years ago in what astronomer Fred Hoyle (now deceased) originally termed, in derision, the *big bang*.

The notion of the big bang was first proposed in 1927 by Belgian astronomer and Catholic priest Georges Lemaître. Well before observations of the cosmic microwave background radiation provided the first good observational support for the theory, Pope Pius XII used the big bang theory to validate Catholic theology. In a speech before the Pontifical Academy, the pope asserted that "creation took place in time, therefore, there is a Creator, therefore God exists."[13] At the urging of academy member Lemaître, however, the pope stopped short of making this an "infallible" pronouncement. Lemaître realized how dangerous that would have been, knowing that his theory like any other was not infallible.[14]

Although a handful of aging astronomers continue to hold out for a steady-state universe,[15] recent observations now make the big bang theory more compelling than ever. Perhaps the strongest confirmation of the big bang came from the Cosmic Microwave Background Explorer (COBE) satellite in 1992. The results of this experiment, and more recent data from newer instruments, have verified the theory of the big bang with a probability approaching certainty.

When the COBE results first became public, mission scientist George Smoot told reporters, "If you're religious, it's like looking at God." The media loved it. One tabloid front page showed the face of Jesus (as interpreted by medieval artists, of course) outlined on a blurry picture of the cosmos.

Ignoring relativity, quantum mechanics, nuclear energy, DNA, and the rest of twentieth-century science, Ross has characterized the COBE satellite results as "the discovery of the century,"[16] and claims they fulfill the prophecies of the Bible. He inflates the importance of the proponents of the steady-state models, like Hoyle, and charges that they are engaged in the ancient assault mounted by science against God. Ignoring the fact that the overwhelming consensus of cosmologists now accept the big bang, Ross would have us think that scientists are involved in some kind of atheist conspiracy to suppress the big bang theory. In fact, steady-state proponents are far from the mainstream. The over-whelming majority of cosmologists today accept the big bang, which has become as fundamental to the framework of cosmology as evolution has become to the framework of biology.

Ross imagines other atheist conspiracies in science. He wants us to believe that Einstein was an outsider who bucked the atheist establishment. Ross refers to Einstein as an "engineer . . . who studied physics in his spare time."[17] Einstein had a Ph.D. in physics, which Ross surely knows about. To call him an engineer is highly misleading, and he certainly did not learn his physics in his "spare time." According to Ross's characterization, Einstein acknowledged "the necessity for a beginning" and "the presence of a superior reasoning power." Ross says Einstein held out "unswervingly, against enormous peer pressure, to belief in a Cre-ator." But, still, Einstein's God was not the personal one of his Jewish heritage but rather the pantheistic god of Spinoza—another name for the order of the universe (see below).[18]

As a matter of fact, Einstein was quite put off by the attempts of many to exploit his authority in support of theism. He said, "It was, of course, a lie what you read about my religious convictions, a lie which is being systematically repeated. I do not believe in a personal God and I have never denied this but have expressed it clearly. If something is in me which can be called religious then it is the unbounded admiration for the structure of the world so far as our science can reveal it."[19] Ross grieves that Einstein did not "live long enough to see the accumulation of scientific evidence for a personal, caring Creator."[20]

Ross says that God does not exist in space-time, but in some dimension beyond both space and time, although he does not tell us how he knows that. The deity supposedly looks down on four-dimensional space-time like the cartoon God who sits on his throne in the clouds looking down on the two-dimensional surface of Earth. Ross's God is not eternal; he is beyond eternity. He is not everyplace; he is beyond place. This is also Ross's answer to the skeptic's ques-tion: Who created the creator? The creator is beyond creation.

Ross dismisses my own arguments[21] about the spontaneous

organization of matter out of the chaos of the big bang as "purely speculative." He, quite incorrectly, insists that "not one example of significant self-generation or self-organization can be found in the entire realm of nature." There are in fact many, from snowflakes to weather patterns. Ross adds, again in contradiction to established physics, "Without causation, nothing happens and without organization by an intelligent being, systems tend to lower and lower levels of complexity." Ross also dismisses Adolf Grünbaum, philosophy's leading living authority on time, asserting that Grünbaum "stumbles over the nature of time."[22]

Stephen Hawking also comes under Ross's critical gaze for the idea, expressed in *A Brief History of Time*,[23] that the laws of physics might eliminate the need for a creator. Ross looks to Rom. 1:19–22 as a basis to assert that "even a brilliant research scientist can waste his or her efforts, in this case on theoretically impossible lines of research, if he or she rejects clear evidence pointing to God."[24]

According to Ross, many prominent theologians have labeled quantum mechanics, the highly successful modern theory of atoms and light, as the greatest contemporary threat to Christianity. However, he tells us that this "Goliath" has been slain by evidence from a recent turn in research which is "sufficient to rule out all theological options but one—the Bible's."[25] He concludes, "The entity who brought the universe into existence must be a Personal Being, for only a person can design with anywhere near this degree of precision."[26]

CUSTOM-MADE FOR LIFE

In her 1998 cover story mentioned in the preface, *Newsweek's* Sharon Begley reported that "physicists have stumbled on signs that the cosmos is custom-made for life and consciousness."[27] Here she refers to the *fine-tuning argument*, which is based on a series of scientific observations called the *anthropic coincidences* that have led some theists to conclude that the universe shows evidence of having been created with us in mind. In his books and radio show, Ross pushes very hard on this theme.

As the argument goes, if the universe had appeared with slight variations in the values of these parameters, it would not have contained human life.[28] Physicists John Barrow and Frank Tipler have collected a large number of examples of these "coincidences."[29] The implications of these findings deserve some close inspection.

In the first place, how likely is our universe, anyway? Very likely. In fact, since we know it exists, the probability is 100 percent! However, we can ask another question: How likely would our universe have been if it were a random selection of all possible universes? Now this would seem like an impossible question to answer. Nevertheless, if we count all the possible configurations of matter in our present universe, we can at least make a rough guess as to the probability that our particular configuration would have resulted if randomly selected. The true probability would be much lower, so this gives a kind of upper limit. Oxford mathematician and cosmologist Roger Penrose has estimated that this probability is one part in 1 followed by 10^{123} decimal places.[30]

The implication seems to be that if our universe had been the result of a toss of the dice, it would have taken many tosses to produce it. Or it might have happened on the first toss. It should be noted, however, that it takes only one toss to produce *some* universe, and who can say that that universe would not have *some* form of life? And who can say that some limit exists on the total number of tosses?

Many theists join Ross in regarding the anthropic coincidences as the long-sought evidence for purposeful design to the universe, although not all are as adamant as he that the evidence is compelling.[31] Believing, like Einstein, that "God does not play dice," they see the universe as specially designed to come out the way it has, with billions of stars and galaxies and intelligent, carbon-based life on at least one planet.

THEISM, DEISM, AND PANTHEISM

In these and other recent developments, we are being presented with novel variations on the ancient *argument from design*. This line of reasoning for the existence of God goes back at least as far as Aristotle (d. 322 B.C.E.), who proposed that the order of the universe implies a series of causes that has to end some place—at the First Cause, Uncaused. In his book *The Nature of the Gods*, the famous Roman lawyer and statesman Cicero (d. 45 B.C.E.) presented the design argument, which he attributed to the Stoics. He also mentioned the alternative position of the atomists that the world was natural and without the need of a creator. The argument from design was incorporated by St. Thomas Aquinas (d. 1274) in two of his "proofs" of the existence of God as the Grand Designer, the First Cause.

Isaac Newton (d. 1727), who was highly skeptical of certain Church teachings such as the Trinity, still thought he saw the need for a creator god to lay down the laws of motion and gravity that he had uncovered. In the clockwork universe implied by Newtonian mechanics, everything that happens is fully determined by natural law once those laws and initial conditions of the universe are put in place. According to this picture, once god created the universe and its laws, being perfect, he needs to step in no further as the clocklike world unwinds according to those laws. Human free will does not exist in the clockwork universe—only god's will. After the creation, everything that happens has already been preordained, written in the book of nature which should be studied alongside scripture to learn god's will.

This theological position, called *deism*, or *natural religion*, with a lowercase *god*, became a popular doctrine during the Age of the Enlightenment that was triggered by scientific developments, in particular, Newtonian physics. However, Newton himself did not hold the position that there was no need for god to act after the creation. In response to queries by theologian Richard Bentley, Newton wrote that he could not explain the particular orientation of planetary orbits or their stability. This left open the possibility that while the elliptical shape of these orbits was determined by natural laws, other properties were determined by God as time went by.

In this manner, Newton allowed a means by which the theology of a continually acting God, a God who intervenes in human affairs, could reconcile itself with science. This avoided the conflict between science and religion that had erupted with the condemnation of Galileo a generation earlier. As we will discuss in detail later, many theologians continue to work at attempting to reconcile God with science.

In the century after Newton, many scientists and other Enlightenment thinkers became deists who saw no role for God after the creation. Several of the Founding Fathers of the American nation, notably Benjamin Franklin, George Washington, John Adams, Ethan Allen, Thomas Jefferson, and Thomas Paine were deists, and indeed, quite anti-Christian. Here are some selected quotations.

> I almost shudder at the thought of alluding to the most fatal example of the abuses of grief which the history of mankind has preserved—the Cross. Consider what calamities that engine of grief has produced!
>
> —John Adams[32]

> Millions of innocent men, women, and children, since the introduction of Christianity, have been burnt, tortured, fined and imprisoned; yet we have not advanced one inch toward uniformity. What has been the effect of coercion? To make one half of the world fools and the other half hypocrites. To support roguery and error all over the Earth.
>
> —Thomas Jefferson[33]

> Religious bondage shackles and debilitates the mind and unfits it for every noble enterprise. . . . During almost fifteen centuries has the legal establishment of Christianity been on trial. What have been its fruits? More or less, in all places, pride and indolence in the clergy; ignorance and servility in laity; in both, superstition, bigotry, and persecution.
>
> —James Madison[34]

The most detestable wickedness, the most horrid cruelties, and the greatest miseries that have afflicted the human race have had their origin in this thing called revelation, or revealed religion.

It has been the most destructive to the peace of man since man began to exist. Among the most detestable villains in history, you could not find one worse than Moses, who gave an order to butcher the boys, to massacre the mothers and then rape the daughters. One of the most horrible atrocities found in the literature of any nation. I would not dishonor my Creator's name by attaching it to this filthy book.

—Thomas Paine[35]

I have found Christian dogma unintelligible. Early in life I absented myself from Christian assemblies.

—Benjamin Franklin[36]

These forcefully expressed opinions belie the claim, commonly heard in today's pulpits and editorial pages, that America was founded as a "Christian nation." The god of deism is most definitely not the Judeo-Christian-Islamic personal God. The latter is the God of *theism*, who answers prayers, allows humans freedom to choose, and is involved in a continuous process of creation. The Founding Fathers quoted above, and others, were not theists and, thus, certainly not Christians. No doubt to the great surprise of most Americans today, the "creator" of the Declaration of Independence, who endowed men with "inalienable rights," was not the Christian but the deist god! And neither God nor Jesus Christ is mentioned once in the Constitution. For good reason. Freedom and democracy are not part of any religious tradition, much less one based on the Bible.

And what is the status of deism today? In 1925 a young German physicist named Werner Heisenberg realized that the new *quantum mechanics* he and others were developing implied that a fundamental indeterminism existed in nature, which contradicted the clockwork universe of deism. Heisenberg's *uncertainty principle* showed that it was impossible to measure both the position and momentum (the product of the body's mass and velocity) of a body simultaneously with unlimited accuracy. Since, according to

Newtonian mechanics, both quantities are needed to predict the future position of the body, the best one could do was compute the probability of finding a body in some region in the future. Thus an element of chance that most likely cannot be eliminated was introduced into physics, with the further implication that everything which happens in the universe is not predestined after all.[37]

Some scientists today, such as Australian physicist, prolific author, and Templeton Prize–winner Paul Davies, have proposed a revised deism in which God designed the world to operate according to both law and chance.[38] This theology, or what one might call "deology," has many attractive features. It allows for human free will, evil in the world, and chance evolution, thus removing much of the logical difficulty that theologians have in reconciling these with the traditional creator God.

However, it takes a mighty feat of apologetics to identify the new deist god with the all-good, all-perfect, all-powerful Judeo-Christian-Islamic God who still takes an active role in the universe today—long after the initial creation. And, even many of those theists who accept evolution cannot abide the implications that arise when one allows a large role for chance in the universe—that reality is ultimately purposeless. They see evolution as the mechanism by which God intentionally produced his most wondrous creation—them.

Both deism and traditional Judeo-Christian-Islamic theism must also be contrasted with *pantheism*, the notion attributed to Baruch de Spinoza (d. 1677) that the deity is associated with the order of nature or the universe itself. This also crudely summarizes the Hindu view and that of many indigenous religions around the world. When modern scientists such as Einstein and Stephen Hawking mention "God" in their writings, this is what they seem to mean: that God *is* Nature.

Einstein's general theory of relativity was criticized in 1921 by Cardinal O'Connor of Boston as "a ghastly apparition of atheism." When Rabbi Herbert Goldstein of the International Synagogue in New York sent Einstein a cablegram bluntly demanding, "Do you believe in God?" Einstein followed his wife's advice and replied, "I believe in Spinoza's God who reveals himself in the orderly harmony of what exists, not in a God who concerns himself with the fates and actions of human beings."[39] In other words, Einstein admitted he did not believe in the God of his heritage; but, being Einstein, he got away with it.

It will be very important to keep in mind that the theist God whose existence is being claimed to be supported by scientific evidence is neither the god of deism nor the god of pantheism. The theist God takes such an active role in ordinary events that he surely must be detectable. And, if not detectable at some reasonable level, then a good case can be made that such a God does not exist.

The pantheist god is beyond the scope of scientific analysis. Rather it is simply a definition, the equating of deity with nature. If, like Einstein, you are an atheist but want to avoid trouble when asked whether you believe in God, you can truthfully say say, "Yes, I believe in god," hoping that no one notices the lower case g and keeping to yourself that you have simply said that you believe in the reality of nature.

The Enlightenment deist god, who is still outside nature, is difficult to verify or rule out since it has stayed out of cosmic and human affairs since the creation, letting its laws do the work. The existence of a creator, in any form, is based on the same design arguments usually given for the Judeo-Christian-Islamic creator God. If they were valid, they would still not provide unique support for theism. If these arguments were shown to be flawed, then parsimony would favor no creator at all, theist or deist.

Even if design in the universe were to be conclusively validated, this would not necessarily imply that the creator was a personal God. A currently inactive, impersonal, creator god remains possible, but that god has nothing to do with the traditional God of theism. Thus, the reason that so many believers give to pollsters for their belief in a personal God, that they see evidence in the design of the universe, does not in fact provide any basis for that particular belief. Unfortunately, they are being misled by theists like Ross into thinking that their perception of design is a reason to believe in the God of theism. Even worse, they are being misled into thinking that their perception of design has a basis in science.

DESIGN DISPUTED

The argument from design has usually been presented on theological and philosophical grounds. Only in the last few years have we seen the debate shift to a scientific (meaning, in this case, empirical) venue. The logic of the argument from design was very effectively refuted by David Hume (d. 1776) in his *Dialogues Concerning Natural Religion*. Hume pointed out that the patterns observed in the world around us could just as well have been organized from within the universe as without.

The argument from design rests on the notion that everything, but God, must come from something. However, once you agree that it is logically possible for an entity to exist which was not itself created, namely God, then that entity can just as well be the universe itself. Indeed, this is a more economical possibility, not requiring the additional hypothesis of a supernatural power outside the universe. As we will see, Hume's argument has only grown stronger with age.

Despite being logically fallacious, the argument from design has remained the most popular justification for belief in a creator. As we saw in the survey described at the beginning of this chapter, some version of the argument from design continues to represent a major rationalization educated people give for their belief. Most are probably unaware of Hume's critique. To them, it is not a matter of logic anyway, but common sense. They see no way that the universe could have just happened, without intent. "How can something come from nothing?" they continue to ask, never wondering how God came from nothing.

In the early nineteenth century, the commonsense design argument found powerful expression in an 1802 treatise by theologian William Paley, *Natural Theology—or Evidences of the Existence and Attributes of the Deity Collected from the Appearances of Nature*. Paley describes a watch being stumbled upon in the heath. Even if you do not know the source of the watch, Paley argues, you would still conclude that was an artifice and not some natural object. He cites the human eye and other organs as obvious examples (to him, at least) of divine contrivance.

Modern preachers have used another variation of Paley's watch: Imagine a hurricane hitting a junkyard and assembling a full Boeing 747 aircraft. "How ridiculous!" they shout, to the delight of their congregations. And yet, they never bother to try to describe the kinds of cosmic winds by which something infinitely more complex—God himself—was assembled.

The rational response to the question of how such intricate complexity as life came to be was first crafted independently by Charles Darwin and Alfred Russel Wallace in 1859, a half-century after Paley. Theirs was the description of a mechanism that made watches and 747s appear as mere child's play. Darwin's and Wallace's focus was on the biological organism. The secret of its mind-boggling complexity lay in natural selection. Today, as we have seen, the evidence for natural selection is overwhelming, and it has become the organizing principle of the science of biology. Dawkins's *The Blind Watchmaker: Why the Evidence of Evolution Reveals a Universe without Design* provides a direct, modern response to Paley, while deftly buttressing David Hume's original argument.[40]

Nevertheless, as we saw in chapter 2, Darwinism is regarded by many theists as the greatest of all threats to their faith. This is because, in their minds, it seems to imply a lack of divine purpose to human life. It has to be wrong, they insist. The universe must have purpose! Life must have meaning! Under the title of "creation science," or, more recently, "intelligent design," these theists continue to battle evolution tooth and nail, claiming equal time for what they see as the legitimate alternative of purposeful, supernatural creation. They assert evidence, but that evidence is not new data gathered to uncover new theories but rather old data molded to fit old preconceptions.

As I have already pointed out, even if evolution by natural selection were totally wrong, this would not, by default, validate the notion of supernatural design. Even without natural selection, other natural possibilities could remain. Life on Earth might have been seeded by an extraterrestrial civilization, as the Raelians assert. True, this would still leave the question of how the extraterrestrials evolved, but this is a natural alternative to a supernatural being for which we would have the same unanswered question. Hume knew nothing about natural selection, or about the possibility of alien civilizations, but he correctly realized that the order of the universe could be, in principle, perfectly natural.

NOTES

1. George Bishop, "Poll Trends: Americans' Belief in God," *Public Opinion Quarterly* 63 (1999): 426.

2. Michael Shermer, *How We Believe: The Search for God in an Age of Science* (New York: W. H. Freeman, 2000), p. 78.

3. Edward J. Larson and Larry Witham, "Leading Scientists Still Reject God," *Nature* 394 (1998): 313.

4. Rodney Stark, Laurence R. Iannaccone, and Roger Finke, "Religion, Science, and Rationality," *American Economic Review Papers and Proceedings* (May 1996): 433–37.

5. Stephen J. Gould, *Rocks of Ages: Science and Religion in the Fullness of Life* (New York : Ballantine Pub. Group, 1999).

6. Owen McCleod, "Science, Religion, and Hyper-Humeanism," *Philo* 4, no. 1 (2001): 68–81.

7. Gould, *Rocks of Ages*, p. 93.

8. Richard Dawkins, "When Religion Steps on Science's Turf," *Free Inquiry* 18, no. 2 (1998): 18–19.

9. Patricia Smith Churchland, *Neurophilosophy: Toward a Unified Science and the Mind-Brain* (Cambridge: MIT Press, 1986); Francis Crick, *The Astonishing Hypothesis: The Scientific Search for the Soul* (New York: Charles Scribner's Sons, 1994); Paul M. Churchland, *The Engine of Reason, the Seat of the Soul* (Cambridge: MIT Press, 1995); George Lakoff and Mark Johnson, *Philosophy in the Flesh: The Embodied Mind and Its Challenges to Western Thought* (New York: BasicBooks, 1999).

10. Ray Kurzwell, *The Age of Spiritual Machines* (New York: Penguin, 2000); Brad Lemley, "Machines That Think," *Discover* (January 2001): 74–79.

11. Hugh Ross: *The Fingerprint of God* (Orange, Calif.: Promise Publishing, 1991); *The Creator and the Cosmos: How the Greatest Scientific Discoveries of the Century Reveal God* (Colorado Springs: NavPress, 1995); *The Genesis Question: Scientific Advances and the Accuracy of Genesis* (Colorado Springs: NavPress, 1998).

12. Ross, *The Creator and the Cosmos*, p. 61.

13. Pius XII, "The Proofs for the Existence of God in the Light of Modern Natural Science," address of the pope to the Pontifical Academy of Sciences, November 22, 1951, reprinted as "Modern Science and the Existence of God," *Catholic Mind* 49 (1972): 182–92.

14. Tony Ortega, "High Priests of Astronomy," *Astronomy* (December 1998).

15. Fred Hoyle, Geoffrey Burbidge, and Jayant V. Narlikar, *A Different Approach to Cosmology: From a Static Universe through the Big Bang towards Reality* (Cambridge: Cambridge University Press, 2000).

16. Ross, *The Creator and the Cosmos*, p. 19.

17. Ibid., p. 52.

18. Ibid., pp. 52–53.

19. Helen Dukas and Banesh Hoffman, eds., *Albert Einstein—The Human Side* (Princeton: Princeton University Press, 1979), p. 42.

20. Ross, *The Creator and the Cosmos*, pp. 52–55.

21. Victor J. Stenger, *Not By Design: The Origin of the Universe* (Amherst, N.Y.: Prometheus Books, 1988).

22. Ross, *The Creator and the Cosmos*, p. 84.

23. Stephen W. Hawking, *A Brief History of Time: From the Big Bang to Black Holes* (New York: Bantam, 1988).

24. Ross, *The Creator and the Cosmos*, pp. 91–92.

25. Ibid., p. 95.

26. Ibid., p. 118.

27. Begley, "Science Finds God," *Newsweek*, July 20, 1998, p. 46.

28. Brandon Carter, "Large Number Coincidences and the Anthropic Principle in Cosmology," in *Confrontation of Cosmological Theory with Astronomical Data*, ed. M. S. Longair (Dordrecht: Reidel, 1974), pp. 291–98; reprinted in *Physical Cosmology and Philosophy*, ed. John Leslie (Amherst, N.Y.: Prometheus Books, 1998), pp. 131–39.

29. John D. Barrow and Frank J. Tipler, *The Anthropic Cosmological Principle* (Oxford: Oxford University Press, 1986).

30. Roger Penrose, *The Emperor's New Mind: Concerning Computers, Minds, and the Laws of Physics* (Oxford: Oxford University Press, 1989), p. 343.

31. Richard Swinburne, "Argument from the Fine-Tuning of the Universe," in *Physical Cosmology and Philosophy*, ed. John Leslie (New York: Macmillan, 1990), pp. 154–73; George Ellis, *Before the Beginning: Cosmology Explained* (London and New York: Boyars/Bowerdean, 1993).

32. John Adams, letter to Thomas Jefferson (undated).

33. Thomas Jefferson, *Notes on the State of Virginia*, 1781–1782, ed.

Merrill D. Peterson (New York: Library of America, Literary Classics of the United States, 1984).

34. James Madison, letter to William Bradford Jr., April 1, 1774.

35. Thomas Paine, *The Age of Reason*, October 25, 1795.

36. Benjamin Franklin, in "Toward the Mystery" (undated).

37. The possibility remains that some subquantum forces exist that account deterministically for what is observed as chance, as in the version of quantum mechanics proposed by David Bohm. See D. Bohm and B. J. Hiley, *The Undivided Universe: An Ontological Interpretation of Quantum Mechanics* (London: Routledge, 1993). However, no evidence has been found to support his proposal, and the theory has serious problems, such as inconsistency with Einstein's special theory of relativity.

38. Paul Davies, *The Mind of God: The Scientific Basis for a Rational World* (New York: Simon and Schuster, 1992).

39. Ronald W. Clark, *Einstein: The Life and Times* (New York: Avon Books, 1971), p. 502.

40. Richard Dawkins, *The Blind Watchmaker: Why the Evidence of Evolution Reveals a Universe without Design* (New York, London: W. W. Norton, 1986).

FOUR

MESSAGES FROM HEAVEN

Nature is what the world would be if there were no God.
—William Dembski, 1999

INTELLIGENT DESIGN

The argument from design stands or falls on whether it can be demonstrated that some aspect of the universe such as its origin or biological life could *not* have come about naturally. The burden of proof is not on the naturalist or evolutionist to rule out supernatural causes. Rather, it is on the supernaturalist to demonstrate that something from outside *nature* must be introduced to explain the data.

This realization finally seems to be hitting home in the community of knowledgeable believers. Although many preachers, lawyers, and politicians still think they can win debating points for religion by poking holes in evolution or otherwise castigating the "dogmatic naturalism" of science, more sophisticated theologians understand that this strategy works only when the debate is held in front of a congregation of faithful parishioners. Similarly, awe-inspiring photographs of galaxies and stars, or detailed maps of the structure of the cosmic microwave background radiation do not signal the existence of a creator—unless those photographs and maps contain some undeniable "message from heaven."

Mathematician and theologian William Dembski says that he has seen just such a message from heaven, written in the patterns of nature. This has made him a hero and a martyr to some, while a fool and a pariah to others.

Dembski is a major spokesperson in the new creationist movement that goes by the designation *intelligent design*, which was briefly introduced in chapter 2. He claims to have *proven* that the order and complexity of the physical world are incapable of being brought about naturally. Thus he directly disputes Hume's case against the argument from design.[1]

Dembski makes his opposition to theists who support evolution very clear:

> *Design theorists are no friends of theistic evolution.* As far as design theorists are concerned, theistic evolution is American evangelicalism's ill-conceived accommodation to Darwinism.[2]

In 1999, after he had gained considerable notoriety from his speeches and writings, Dembski was invited by President Robert Sloan of Baylor University, a moderate Baptist institution in Waco, Texas, to become director of a new institute for intelligent design studies on campus. The center was to be named after chemist Michael Polanyi, who questioned that the world could be explained by natural laws alone.

The Polanyi Center was established by executive order, without the normal faculty consultation, and when the Baylor faculty heard about it, many were unhappy. Charles Weaver, a professor of psychology and neuroscience at Baylor, typified the concern of professors about the reputation of the institution: "When you say Baylor now, people are going to go, 'Oh, yeah, they have that creationist center.'" Waco also happens to be the unfortunate location of the Branch Davidian compound that was destroyed with great loss of life during a federal assault in 1993. This added to the concern of the faculty that Baylor, really a very fine university, might become associated with kooks.[3]

The Baylor faculty senate voted twenty-seven-to-two to dismantle the new center, but President Sloan refused to comply. Instead, he invited an outside committee of scholars to consider the legitimacy of the center and evaluate intelligent design as an academic discipline. Despite Dembski's vigorous objections that the committee was not qualified to provide peer review of his work, it recommended that this work be allowed to continue. The committee report was conciliatory. It said that it considers "research on the logical structure of mathematical arguments for intelligent design to have a legitimate claim to a place in the current discussions of the relations of religion and science." They recommended that the Polanyi name be removed and that an advisory committee oversee the work of the center. Perhaps most significantly, the committee specified that the center was more naturally at home within the Baylor Institute of Faith and Learning and that it would be too restrictive for it to focus attention "on a single theme only, such as the design inference."[4]

Dembski nevertheless saw the decision as a great victory in the battle of good against evil. In a press release, he exulted, "Dogmatic opponents of design who demanded the center be shut down have met their Waterloo. Baylor University is to be commended for remaining strong in the face of intolerant assaults on freedom of thought and expression." After refusing to retract this incautious statement, Dembski was removed as center director by the Baylor administration.[5]

Dembski was replaced as director of the center, now called the Baylor Project on Religion and Science, by Bruce Gordon. Gordon has written that one of the principle reasons for the academic resistance to design-theoretic research is that it has been "hijacked as part of a larger cultural and political movement." He adds, "Inclusion of design theory as part of the standard discourse of the scientific community, if it ever happens, will be the result of a long and difficult process of quality research and publication. It also will be the result of overcoming the stigma that has become attached to design research because of the antievolutionary diatribes of some of its proponents on the one hand and its appropriation for the purpose of Christian apologetics on the other."[6]

As these events once again illustrate, the dispute between creationists and scientists is more political than academic, and the battle metaphor used by Dembski was not inappropriate even if it was ill-advised. We saw in chapter 2 how law professor Phillip Johnson is leading a political war against what he perceives as a dogmatic bias toward naturalism in mainstream science. Dembski's "battle of Waterloo" has by no means been fought to completion, and it is far from clear which side will emerge victorious.

While numerous expert critiques of Dembski's ideas exist in the literature,[7] I wish to focus on some glaring errors that cast grave doubt on his whole program.

INFORMATION THEORY

In his book *Intelligent Design*, Dembski makes the assertion that "intelligent causes are necessary to explain the complex, information-rich structure of biology and that these causes are empirically detectable."[8] This is a scientific statement and so should be evaluated scientifically, outside the political arena.

Dembski has become prominent for his claim to have brought modern information theory to bear on the issue of design. However, his writing on the subject is ambiguous, and considerable debate exists on exactly what he has done. What is clear is that he has not applied information theory as it is conventionally practiced in the field. In order to be as precise as possible in justifying this statement, I must get technical at this point. Those readers who have some knowledge of mathematics should be able to follow the arguments. For those who do not, I will also summarize my conclusions in the less-precise medium of words.

Dembski defines the "measure of information in an event of probability p" as

$$I_D = -\log_2 p \qquad (1)$$

where the base-2 logarithm is used.[9] He cites as a reference *The Mathematical Theory of Communication* by Claude Shannon and Warren Weaver.[10]

Shannon is regarded as the father of information theory. Working for Bell Labs, a major private research company in the phone industry (now reorganized), Shannon was concerned with the efficient communication of electronic signals. Let me give a short summary of his theory, which is now widely applied in communications and computer engineering.

Suppose we want to transmit a message containing a single symbol, such as a letter or number, from a set of n symbols. Shannon defined a quantity

$$H = -\Sigma_i\, p_i \log_2 p_i = -< \log_2 p_i > \qquad (2)$$

that he called "the entropy of the set of probabilities $p_1 \ldots p_n$" for the symbols in the message. That is, p_i is the probability for the i'th symbol in the list. The sum over i goes from 1 to n. Because of the base-2 logarithm, the units of H are *bits*.

The angle brackets in (2) refer to the average of the enclosed quantity, and the fact that H is an average over an ensemble of symbols is important to keep in mind in the ensuing discussion.

In today's literature on information theory, H is called the *Shannon uncertainty*. The information, R, carried by a message is defined as the decrease in Shannon uncertainty when the message is transmitted. That is,

$$R = H(\text{before}) - H(\text{after}) \qquad (3)$$

If we consider the special case when all the probabilities p_i are equal, we get the simpler form

$$H = -\log_2 p_i \tag{4}$$

Let me illustrate the idea of information with a simple example of a single-character message that can be one of the eight letters S, T, U, V, W, X, Y, or Z with equal probability. Before the message is transmitted, $n = 8$, $p_i = 1/8$, and $H(\text{before}) = -\log_2(1/8) = \log_2(8) = 3$. After the message is successfully transmitted, we know what the character is, so $p_i = 1$ and $H(\text{after}) = -\log_2(1) = 0$. Thus $R = 3$ bits of information are received as the uncertainty is reduced by 3 bits.

Now suppose that the message is a little garbled so that we know the symbol transmitted is either a U or a V, but we cannot tell which, and they have equal probability. Then, after the message is received, $p_i = 1/2$ and $H(\text{after}) = -\log_2(1/2) = 1$. In that case, $R = 3 - 1 = 2$ bits of information are received.

Dembski's definition of information, $I_D = -\log_2 p$, is of the same form as the Shannon uncertainty in the special case of equal probabilities, as given in (4). However, we can immediately see that this definition is not conventional and will equal R, as given in (2), only for equal probabilities and when the transmission is perfect so that $H(\text{after}) = 0$. While Dembski refers to Shannon, he does not mathematically derive the expression for information he uses from Shannon's expression, nor justify it by any other method. His examples, however, indicate that he does not limit himself to equal probabilities within an ensemble of symbols or "events." Neither does he average over the ensemble. In fact, his so-called information is really just another way of writing the probability, p, of an event in logarithmic form. This quantity is called *surprisal* in the literature, although too trivial to warrant a separate term.[11]

Before continuing with Dembski's theory, let us take a closer look at the interpretation of the Shannon uncertainty, H. Shannon notes that "the form of H will be recognized as that of entropy as defined in certain formulations of statistical mechanics," referring to the classic monograph *The Principles of Statistical Mechanics* by

Richard Tolman,[12] which has been on my bookshelf since my graduate student days. Shannon explicitly states that "H is then, for example, the H in Boltzmann's H-theorem." Actually, in statistical mechanics the quantity I will call H_{SM} is defined without the minus sign and using the natural logarithm:

$$H_{SM} = \Sigma_i \, p_i \log_e p_i \tag{5}$$

However, Shannon notes that any constant multiplying factor, positive or negative, could have suited his purposes, since it just sets the units. As we saw above, the choice he made in (1) gives units in bits.

Boltzmann's H-theorem is one of the most famous in physics. In the 1890s, Ludwig Boltzmann showed that no matter what the initial distribution of velocities of the molecules of a gas, the effect of collisions led to a decrease in H_{SM} with time until a minimum is reached, at which point the gas is in equilibrium.[13]

Boltzmann and Josiah Willard Gibbs found that the laws of classical thermodynamics could be derived from statistical mechanics on the assumption that matter was composed of atoms, whose existence had not yet been fully confirmed by experiment at that time. In particular, the quantity H_{SM} was seen to be simply related to the thermodynamic entropy S by $S = -kH_{SM}$, where k is Boltzmann's constant. The H-theorem then implied that the entropy approaches maximum at equilibrium, as was well known from thermodynamics. This gave a statistical explanation for the second law of thermodynamics, which says that the entropy of an isolated system will increase with time or stay constant, that constant being achieved when equilibrium is reached.

The relationship between the entropy, S, of statistical mechanics and Shannon's uncertainty, H, then is,

$$S = k \log_e(2)H \tag{6}$$

That is, they are equal within a constant and have the same sign. So Shannon was justified in calling H the "entropy."

Summarizing the conclusions of this section: (1) Dembski's definition of information is not that used in the discipline of information theory, (2) information is conventionally defined as the change of a quantity called the Shannon uncertainty, and (3) entropy and Shannon uncertainty are equal within a constant.

CONSERVATION OF INFORMATION

In his book *Intelligent Design* and in other writings, Dembski claims to prove a principle he calls, borrowing the term from 1960 Nobel laureate in medicine Peter Medawar, the *law of conservation of information*. According to Dembski's—but not Medawar's—version of this principle, the number of bits of information cannot change in any natural process such as chance or the operation of some physical law. As Dembski states it, "Chance and law working in tandem cannot generate information."[14] I will try to show that this is incorrect, when interpreted as some universal principle applying under all circumstances, which Dembski seems to do.

The basic idea of conservation of information is simple, and is illustrated in figure 4.1. Suppose we start out with a certain number of bits of information about a system. Let the system be composed of five coins. Any configuration of heads (H) or tails (T) is information that can be represented by five bits. For example, HTTHT = 10010. According to Dembski, two possible natural processes can act on that information. One is some well-defined operation that can be likened to the action of a physical law or computer algorithm. For example, the operation might be: every time a flipped coin lands on the table, turn it over. Thus we have HTTHT → THHTH, or 10010 → 01101. Clearly, the number of bits has not changed and so, while the message may be different in content, it contains no more or less information than previously. In this process, at least, information is conserved.

In his book *The Limits of Science*, Medawar described the impossibility of creating new information from closed logical systems: "No process of logical reasoning—no mere act of mind or com-

"Proof" of Conservation of Information

Law just rearranges bits by some rule

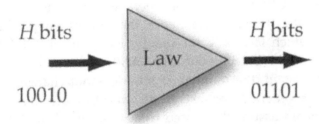

Chance just scrambles bits randomly

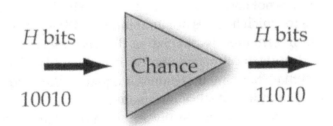

Conclusion: The generation of information cannot be natural.

Fig. 4.1. A simplified proof of "conservation of information." Natural processes of law or chance just rearrange bits of informaiton but do not generate new bits.

puter-programmable operation—can enlarge the information content of the axioms and premises or observation statements from which it proceeds."[15] However, unlike Dembski, Medawar did not claim this as a universal principle but instead claimed that it applied only to *closed* systems. Medawar also made no claim that the same rule applied to chance processes, which Dembski includes in his version of the principle.

Let us look at what happens to our original information representing five coin tosses under the operation of random chance. Regardless of the original bit sequence, the process will produce a new one in which H and T are equally likely. Thus HTTHT → TTHTT or HHTTH (10010 → 00100 or 11001) or any other possible permutation. Again, the number of bits does not change, so no information is generated or lost. Notice that Dembski has, as he did above, again assumed a closed process in which no bits can flow in or out. This is not the most general situation.

Thus, Dembski claims, the very existence of information in the universe is irrefutable evidence for the existence of design. This conclusion has left many people impressed. For example, Rob Koons, an associate professor of philosophy at the University of Texas, calls Dembski the "Isaac Newton of information theory."[16]

Hardly. First, as I have already noted, Dembski's definition of information does not correspond to that used in the field, except as a special case to which he does not limit himself. We have seen that conventional information is defined as the decrease in Shannon uncertainty during the transmission of a message. Furthermore, we have seen that Shannon uncertainty is equal, within a constant, to the entropy used in statistical mechanics.

As has been well known in physics for more than a century, entropy is *not* conserved. In fact, the second law of thermodynamics says that the total entropy of an isolated system of many bodies must remain constant *or increase,* as implied by Boltzmann's *H*-theorem, discussed above. Entropy can even decrease for nonisolated systems, which happens when they are organized from the outside, or in any system with small numbers of particles. In fact, in the example I gave of the transmission of a message, the entropy/uncertainty is seen to decrease. This is an illustration of a nonisolated system, the transmitter, sending information to another nonisolated system, the receiver.

Indeed, every time you rub your hands together, you are making entropy. From an information standpoint, uncertainty is increasing (the molecular motions in your hands are becoming more irregular) and so information is being lost. It is possible to come up with many examples in which information is not conserved. Thus, Dembski's "proof" fails because it violates the second law of thermodynamics. Actually, Dembski is aware that information can degrade,[17] but this only demonstrates his inconsistency. His "law" of conservation of information does not permit this.

Let us move from individual systems to the universe as a whole. Under the naturalistic assumption that no forces act on it from the outside, the universe is an isolated system. Thus, it would appear that no gain of conventional information (negative entropy) with time is possible in the universe as a whole. However, the universe could have begun in complete disorder without violating the second law, and local pockets of order can still form as time progresses. This is made possible by the expansion of the universe, which continually opens up more room for order to form (see chapter 6 for further discussion of this point).

Now, perhaps Dembski might argue that it was not his intent to define information in terms of the Shannon uncertainty, although he uses Shannon as a reference and mentions no other source. In any case, we can show that we can still imagine natural processes adding bits of Dembski information to a system. A computer simulation can be used to illustrate this, but the creationists can then say that it is still "designed." Let us consider an example that involves only chance, with no designer intervention.

Suppose we have two bar magnets, one sitting on top of the other, as shown in figure 4.2(a). Because of their mutual attraction, only the two configurations shown, with either both north poles up or both south poles up, will be stable. This can be specified by one bit of information, say $H = 1$ for north poles up and $H = 0$ for both north poles down. Note that $H = I_D$ in this case.

We open the window and a random breeze comes through and knocks the magnets apart. Assume they are constrained so they cannot fall on their sides but must always land vertically. Now, because the poles are no longer in contact, the four configurations shown in (b) are possible. We then need two bits to describe the situation: $H = 11$ for both north poles up, $H = 10$ for the first north up and the second down, $H = 01$ for the second north up and the first down, and $H = 00$ for both north poles down. Thus, the information in the system has increased by one bit as the result of a chance process. (We would need even more bits to describe the possible orientations for the magnets on their sides.) Thus information is generated by chance, in violation of Dembski's law of conservation of information. Such a law cannot be found in the standard usages and practices within the field of information theory.

PANNING FOR DESIGN

The law of conservation of information is not the only unloaded weapon to be found in the arsenal of intelligent design. In tune with the arguments of Ross presented in chapter 3, Dembski attempts to show that design is evident by virtue of what, in his personal estimation, are probabilities that are too low for the natural production of order.

In this approach, Dembski seems to be utilizing his own definition of information, which we saw in the previous section is not the conventional one. In fact, he is simply doing a trivial mathematical transformation on the probability of an event, converting it to bits by taking the negative of its base-2 logarithm. In Dembski's usage, the lower the probability of an event, the greater the information contained in that event.

Dembski introduces a series of "filters," which he applies to observed phenomena in order to determine whether or not they are designed. He tests these filters by also applying them to examples of human design, on the assumption that supernatural design must follow the same rules. His filters pass only information that

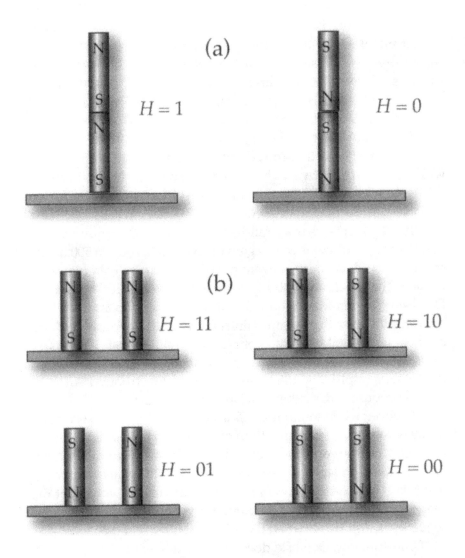

Fig. 4.2. Example of the generation of information by chance. The bar magnets in (a) sit in a weakly stable state that requires only one bit to describe, north poles up or north poles down. In (b), a random breeze has knocked the magnets free, and they can land in the four possible states shown that require two bits to describe. Thus chance has generated one bit of information, in violation of Dembski's law of conservation of information.

is both *complex* and *specified*. The resulting *complex specified information* is then interpreted as the consequence of intelligent design.

Let me illustrate these criteria in terms of the example of five consecutive coin tosses used earlier. We saw that this system has five bits of information. Suppose that, before the first toss, we *specify* a particular sequence, say all heads, HHHHH. Or it could be all tails or any other sequence, such as HTTHT, as long as it is specified in advance.

Now, five heads in a row, or any other sequence of five coins, will happen frequently by chance. On average, about one in every thirty-two tosses of 5 coins will land with all 5 heads up. However, suppose we do the experiment with 500 coins instead of 5 and specify in advance that all fall heads up. It would require $2^{500} = 10^{150}$ tosses of 500 coins each, again on average, to obtain 500 heads specified in advance in an event of 500 consecutive coin tosses by chance. That is, the probability for this outcome is 10^{-150}, and the Dembski information contained in the event is 500 bits.

Dembski says this is impossible, for all practical purposes, and defines as "complex" any event containing at least 500 bits of information.[18] This, he notes, is a far more stringent bound than the 166 bits implied by the "universal probability bound" of $10^{-50} = 2^{-166}$ proposed by mathematician Emile Borel.[19]

However, while some *prespecified* sequence of 500 bits, such as all heads or any other specific pattern of heads and tails selected before the fact, has this very low probability of being chance, the probability for *any* sequence of heads and tails in 500 or any number of tosses is 100 percent! So, if after the coins are tossed we look at the sequence that is produced, we cannot very well say that particular sequence is impossible when there it is, staring us in the face.

Unfortunately, Dembski does not define specificity as precisely as he does his albeit idiosyncratic definition of complexity. In the coin example I have used, the sequence is specified in advance. This is fine. However, Dembski cannot leave it at that, because then his whole program to detect design *after the fact* would be defeated. So, as a dubious and dangerous tactic, he allows specificity to be postdetermined. This approach is presumed to capture "design information" that would otherwise be written off to chance, perhaps rightly so.

Although specificity is difficult to define, like pornography you are supposed to know it when you see it. Dembski uses an example from the film *Contact*, based on Carl Sagan's novel, in which the character played by actress Jodie Foster detects a signal from outer space containing the sequence of prime numbers up to 101. She concludes that it comes from an intelligent source. Indeed, if such a message were to come from the skies, then we would all become believers that something out there knows mathematics. However, the message that Dembski claims to detect when he applies his filters to physical systems like biological organisms is not comparable to the sequence of prime numbers up to 101.

For example, he claims that "the CSI [complex specified information] of a flagellum far exceeds 500 bits."[20] What he is really saying is that the probability of a flagellum being produced by natural processes is less than 10^{-150}. He does not know that. He simply asserts it as a fact. As Dawkins and others have shown, natural selection is fully capable of generating large amounts of information. Furthermore, the huge amount of apparently random and useless DNA sequences in the functioning genomes of organisms indicates that a strong element of chance was involved in their development. Living structures do not at all resemble Paley's watch or a Boeing 747 aircraft. In fact, the genetic blueprints of living structures show far more evidence of random processes than they show evidence of effective adaptation by the blind processes of "design" by natural selection, and none for intelligent design.

Physicist, and theist, Howard Van Till points out that Dembski's definition of complexity is highly unorthodox when you look at how it is applied. For example, Dembski argues that any biotic system is complex if the probability for it being assembled by natural processes is less than 10^{-150}. This makes complexity not a property of the system but the means by which it is actualized. [21]

Furthermore, when Dembski actually calculates the probability for a specific system, like the flagella in *E. coli* bacteria, he does so by assuming that the system was assembled by chance processes alone. Van Till comments: "We reject that argument as a totally unrealistic caricature of how the flagellum is actualized and an approach that totally ignores the role of the bacterial genome in coding for all of the structures and functions that contribute to the nature of *E. Coli*."[22]

NO FREE LUNCH

Dembski's latest effort to demonstrate the inability of natural processes to solely account for the evolution of life involves computer programs that attempt to solve optimization problems.[23] As a simple example of an optimization problem, suppose you are hiking in the Rocky Mountains and wish to find the best path to take to reach the bottom of a certain valley, where the best path may not be the shortest one but that requiring the least expenditure of energy. In this case, the energy is an example of a *cost function* or *fitness function*. The search algorithm would try to find the path that has minimum total expended energy. A contour map of the Rockies would constitute an example of what is referred to as a *fitness landscape*, with mountain peaks requiring high cost to scale and valleys requiring low cost.

The simplest optimization method is to generate a large number of random paths and choose the best from the set. For any given problem, this can generally be improved upon by a number of techniques, such as examining the cost function around the current position and moving in the direction where it is lowest. In my example, you would walk downhill from where you are to a new position, look around again, and continue downhill. This may get you down to the desired valley quickly with minimum expenditure of energy. But it may also bring you down into a different valley than the one you are seeking and confront you with mountains to climb or force you to take a long path around to seek out a path that will get you through them. The pioneers of the American West were faced with such problems, and had no computers to guide them.

In a certain subclass of optimization programs, the programmer does not specify the algorithms in detail but uses methods analogous to those of natural selection in which different solutions are tried and those aspects that work better than others are retained while the less efficient aspects are discarded. These are called *evolutionary algorithms*. For example, the program might find that going downhill all the time is bad, while occasionally moving in a random direction instead is better. So, without guidance from the programmer, it may settle on a solution that combines the two techniques in some ratio.

It turns out that for the general class of search algorithms, not just evolutionary ones, no universal algorithm exists that will work for all problems. Certain "no free lunch" theorems have been proven which show that the performance of an algorithm when averaged over all cost functions performs no better and no worse than any other algorithm, including blind search.[24] What works best for the Rockies may not work for Nebraska, and when all landscapes are considered, nothing works better than chance.

Dembski uses this result to assert that Darwinian mechanisms are therefore no better than chance. Since, he claims, chance alone cannot produce the complexity of life, it follows that natural selection cannot be the sole mechanism for biological evolution.

Richard Wein has provided an exhaustive review of *No Free Lunch* and Dembski's earlier work.[25] He points out that the type of search algorithms addressed by the no-free-lunch theorems, including evolutionary algorithms, do not apply to biological evolution. While evolutionary algorithms use natural selection, they do not and are not meant to simulate biological evolution. Biological organisms do not have fixed-cost functions but ones that change over time in response to changes in the population.

H. Alan Orr has seen similar flaws in Dembski's reasoning. Biological evolution has no preset target, no specific fitness landscapes like the Rocky Mountains. As Orr puts it: "Darwinism, I regret to report, is sheer cold demographics. Darwinism says that my sequence has more kids than your sequence and so my sequence gets common and yours gets rare. If there's another sequence out there that has more kids than mine, it'll displace me.

But there's no preset target in this game. (Why would evolution care about a preset place? Are we to believe that evolution is just inordinately fond of ATGGAGGCAGT . . . ?)."[26]

In another review, Jason Rodenhouse agrees that fitness landscapes evolve along with organisms, calling this "a bedrock principle of modern ecology."[27] Jeffrey Shallit points out that Dembski has avoided discussing the types of computer programs that deal with *artificial life*, which more closely relate to biological evolution: "The field of artificial life evidently poses a significant challenge to Dembski's claims about the failure of evolutionary algorithms to generate complexity. Indeed, artificial life researchers regularly find their simulations of evolution producing the sorts of novelties and increased complexity that Dembski claims are impossible."[28]

IRREDUCIBLE COMPLEXITY

In assigning a low probability for natural selection, Dembski relies partially on the work of his close colleague, biochemist Michael Behe. The publication of *Darwin's Black Box: The Biochemical Challenge to Evolution* quickly launched Behe into the constellation of intelligent design stars, right alongside Dembski. In this book, Behe introduces the notion of *irreducible complexity*. A system is irreducibly complex when it is "composed of several well-matched, interacting parts that contribute to the basic function, wherein the removal of any one of the parts causes the system to effectively cease functioning."

Behe applies this to various biological systems such as blood-clotting and cellular flagella, but the idea is most simply explained with the prosaic example of a mousetrap.[29] Remove any single element, say the hammer or spring, and the mousetrap ceases to function. Like Paley's watch, the mousetrap is thus an example of an artifact that cannot have arisen by natural selection. As Behe explains it, "In order to be a candidate for natural selection, a system must have *minimal function*: the ability to accomplish a task in physically realistic circumstances."[30]

Behe insists that biology contains many such minimal systems which could not have evolved by natural selection. "An irreducibly complex system," he says, "cannot be produced directly (that is, by continuously improving the initial function, which continues to work by the same mechanism) by slight, successive modifications of a precursor system, because any precursor to an irreducibly complex system that is missing a part is by definition nonfunctional."[31]

When he wrote his book, Behe apparently was unaware that six decades earlier, Nobel Prize–winner H. J. Muller had already considered irreducibly complex systems, although he did not use the term, and had shown in some detail how they could have evolved naturally.[32]

Behe's assertions have been widely refuted.[33] Critics point to at least one fundamental mistake: Behe assumes that each part of a biological system always had the function it now performs. In fact, countless examples of changing function exist. Thus, a part evolves by natural selection by virtue of one function, and then is gradually adapted to another function as the larger system evolves. Even Behe's mousetrap parts were not invented solely for a mousetrap. Springs and hammers had other functions long before being utilized in rodent extermination.

Like Dembski and other creationists who fail to do their homework, Behe only damages his own credibility with assertions that are provably incorrect. For example, he claims that

> there is no publication in the scientific literature—in prestigious journals, specialty journals, or books—that describes how molecular evolution of any real, complex biochemical system either did occur or even might have occurred.[34]

Not really. Biologist Kenneth Miller is one of the new breed of theistic scientists who think they can find a place for God within the framework of evolution. He has checked Behe's claim for himself. Taking what he describes as a "quick look" at the literature, Miller easily found "four glittering examples of what Behe claimed would never be found."[35]

Chemist Behe does not make a sufficiently strong case for intelligent design in biological systems for Dembski to be justified in his assigning a low probability for natural selection in his design filter. Far from "proving" the need for intelligent design, Dembski's design filter amounts to nothing more than inventing probability estimates after the fact so that the answer comes out favorable to his thesis of intelligent design.

Furthermore, even if one computes a low probability for an event arising from natural processes, this does not prove it cannot have happened that way. Low-probability events occur naturally every day. People win lotteries, are struck by lightning, and run into long-lost relatives. One can infer design as the more likely alternative than random chance only by also computing the probability for design as well as chance and then comparing the probabilities for the two alternatives.

The argument from design, in both its historical and contemporary manifestations, is based on observations and so I have treated it as a scientific matter. However, we have seen that whether the data come from cosmology, biology, or everyday human experience, their interpretation, in terms of the origins of the design seen in them, is in the eye of the beholder. All the reasons to believe based on design reduce to a single reason—a human propensity to see deep and meaningful patterns in the world around them, even when these patterns are not present.

In the United States the new creationist movement has convinced many people and their political servants that scientists are being unfair in not supporting the teaching of alternatives to evolution in science classes. They say it is censorship to exclude intelligent design from those classes. The usual argument raised against teaching intelligent design is that it unconstitutionally promotes religion. Design promoters, however, insist that they have no particular designer in mind. No one believes them, but skilled lawyers arguing for the cause of impartiality on their behalf could probably prevail in court. In any case, a better argument exists: Intelligent design theory, as currently formulated by its leading proponents, should not be taught in science classes because it is provably wrong.

THE ROLE OF RANDOMNESS
IN GENERATING COMPLEXITY

Although the specific arguments that Dembski, Behe, and other design theorists have presented so far are fatally flawed, they still represent the deep intuition of most people that the complex structures we see around us cannot have come about naturally. And no structure seems more complex and more difficult to understand than the human brain. It seems clear to many people that the human brain itself is evidence for design. Surely, they suppose, the properties we label as mind—thinking, feeling, creating—are more than the action of a collection of atoms.

In *The Emperor's New Mind, Shadows of the Mind*, and other writings, mathematician Roger Penrose claimed to show that the brain was not simply a computer, at least, as we understand a computer to be.[36] On the basis of certain, sophisticated incompleteness theorems that have been derived in mathematics and computer science, Penrose argued humanlike "artificial intelligence" was impossible for any physical system operating according to computable algorithms based on known physics. Of course, the architecture of the brain is quite different from today's commercial computers. However, Penrose insists that his conclusions also apply to neural networks such as the brain. He says that something more than computation is happening in the brain.

Penrose admits that he is a Platonist who regards proven mathematical truths as more real than the concrete objects of our experience. He argues that some mathematical truths cannot be established by the execution of a computer algorithm and so something more than computation is going on in a mathematician's brain. Penrose does not think that this something is supernatural. Rather, he seeks a physical mechanism not present in computers and suggests that it might be found in quantum gravity.

I have argued elsewhere that quantum effects of any sort, much less quantum gravity, are unlikely to play a major, direct role in human thinking.[37] Of course, quantum effects predominate within the atoms and molecules of the brain, but the basic physics and chemistry of biological systems are not uniquely different from those of inanimate systems like rocks. A carbon atom in a rock has the same energy levels as a carbon atom in a brain cell. The issue is whether additional, larger-scale quantum effects are present in the brain.

While biological structures, in the brain or elsewhere, are "microscopic" in the sense that they are not visible to the unaided human eye, they are large on the quantum scale and are still well described by classical mechanics. In a concrete example, the mass, m, of the vesicle that carries transmitter chemicals across a synaptic gap is typically 10^{-22} kilogram, a hundred thousand times more massive than a hydrogen atom. The speed of the vesicle, v, is on the order of 10 meters per second, so its momentum, $p = mv$, will be on the order of 10^{-21} kilogram-meters per second. The synaptic gap distance, d, is about 10^{-9} meter, ten times the diameter of an atom. This is to be compared with the de Broglie wavelength of the vesicle, $\lambda = h/p$, where h is Planck's constant, which is on the order of 10^{-12} meter. Quantum interference effects would occur if the vesicle's de Broglie wavelength were comparable to or larger than the gap distance. In fact, it is three orders of magnitude smaller. This difference in size scale places even the tiniest of functional cellular structures well within the scope of classical mechanical description.

Of course, quantum effects such as superconductivity can occur at even macroscopic distances, especially for low-mass particles such as electrons cooled to very low temperatures where their momenta are very low and their de Broglie wavelengths correspondingly high. But, so far, no analogous processes have been discovered in the brain.[38]

Nevertheless, a simple mechanism exists, which is well known to complexity theorists, that can enable the brain or an electronic circuit to act in a noncomputable way and perform creative acts. This process involves random fluctuations. Fluctuations in brain currents can be expected to be induced by external sources in the environment such as cosmic rays, internal sources such as radioactive potassium in blood, or just thermal motion. Like the fluctuations that provide for mutations in the evolutionary process, these might serve to trigger what complexity theorists call a *bifurcation*, when a system moves from one quasi-stable state to another.

Suppose, for example, you have a computer program that is searching for the solution to a problem. You would like to avoid the situation in which the search continues over and over again using the same rules. By nudging the system to another state, it might find the solution by applying a slightly different set of rules. Thus it becomes less rigid, more "creative." Game theorists have long known that intentionally introducing random shifts can be a good gaming strategy.

While Penrose argues that such a process would still be computable,[39] a simple example demonstrates how randomness can break open the bottleneck that can occur in a strictly computable process.[40] In this example, a cop is chasing a robber who can move freely between two houses. The two start off in different houses and have the choice of changing houses or staying put. The robber is caught if he ends up in the same house as the cop. If the cop conducts her search following instructions in the police manual, "by the book," and the robber has a copy of that manual, then the robber will never get caught. He knows the algorithm, so he can move into the other house every time the cop decides to switch. However, if the cop breaks the rules and makes a completely random choice on whether to move into the other house or not, the robber can only guess and will eventually be caught.[41] This illustrates what so many people find so deeply counterintuitive—that chance can in fact produce a desired creative outcome which may be impossible by any other means.

Regardless of the true extent to which randomness plays a role in the brain, what Penrose, Dembski, and others have failed to acknowledge is that it can provide just the push a complex system needs to generate a higher level of complexity. This is precisely the role chance plays in evolution, and, because it is unplanned and undesigned, it is an anathema to those who desperately need to see cosmic purpose in their lives.

Philosopher David Roche has pointed out that Dembski confuses two concepts in his theory of intelligent design: *information* and *complexity*.[42] As I have indicated above, information is degraded by natural processes, in violation of Dembski's law of conservation of information but demanded by the (well-confirmed) second law of thermodynamics. Complexity, however, can increase by a combination of randomness and selection.

As physicist Taner Edis explains it: "A random function, because it is patternless, can be used to break out of any predefined framework. It serves as a novelty generator."[43]

Roche adds:

> The great achievement of Darwinism is not that it explains the origins of information (in the Shannon sense), but that it explains the origin of complexity. And, it does so in terms of a completely material process: *random* mutation followed by *nonrandom* selection. Via such a process, the simple can give rise to the complex; "from so simple a beginning endless forms most beautiful and most wonderful have been, and are being, evolved."[44]

Recently Dembski has seemed to tire of attempting to answer his many scientific critics, a huge task by any measure. At his request, I provided him with a copy of this chapter and chapter 2, inviting comments and corrections, but he never responded. Invited to speak at a conference of skeptics in California in June 2002, Dembski almost completely ignored the important areas of dispute and wished his audience "good luck" in convincing the mass of people of the validity of evolution. He referred to polls showing that the majority of Americans believe in creationism and argued that intelligent design will win, not because it is true but because most people believe it to be true.[45]

THE GOD OF THE GAPS

People have always looked to magic to explain what they do not understand. In their early days, humans knew little and so magic was everywhere. As science developed, providing "natural" explanations for many phenomena, the need for magic diminished. By the seventeenth century and the rise of Newtonian physics, it seemed that very little room was left for magic.

However, as we saw earlier, Newton himself conceded that his laws of mechanics and gravity did not account for everything. For example, while the elliptical orbits of planets followed from his laws of mechanics and gravity, those laws did not determine the particular orientations of planetary orbits. This left an opening for a little magic in the arrangement of the planets.

Thus originated the modern version of the doctrine of the *God of the gaps* in which the creator leaves holes in his own laws to allow for human free will and where he pokes his finger occasionally to make sure that events still follow along according to his divine plan. This implies that God can be seen in the unexplained features of natural law, where his miraculous intervention should be evident.

As time progressed, however, the gaps in Newtonian physics gradually narrowed, nonmiraculously, leaving smaller and smaller holes where God could poke his finger. Pierre Simon Laplace (d. 1827), who generalized mechanics and placed it on an elegant mathematical footing, believed all the gaps were filled by the *Newtonian world machine*.

Still, the God-of-the-gaps doctrine has resurfaced often over the centuries, especially in regard to those sciences not as highly developed as Newtonian physics. This doctrine is experiencing another revival today. Breaches in physics, cosmology, and biology, real and imagined, are being proposed as places for God's purposeful action.

By itself, the existence of gaps in science is not unusual. Unexplained features often appear during the development of scientific theories, but are usually repaired by further developments. This has been the normal course of science. However, in recent times assertions are being made that certain cracks in the traditional, purely material, edifice of science cannot be repaired by the application of additional, conventional cement. Rather, a concoction of ephemeral spirits must be added into the mix.

Back in the early nineteenth century, geologists uncovered stratification and other features that indicated great upheavals in Earth's crust in the past. The causes of these gigantic movements were not immediately evident. In 1819 Oxford geologist William Buckland announced that these observations confirmed scripture and were the consequence of God's actions. He asserted that science can only collect data, that revelation was needed to understand the ultimately divine causes of phenomena.

The eminent Victorian geologist Charles Lyell undermined Buckland's claim by providing plausible natural explanations for geological changes. Still, Lyell saw God's intervention as necessary to explain the existence of life on Earth. Closing Buckland's gap, Lyell left another open. However, Charles Darwin, who, took Lyell's *Principles of Geology* with him on the voyage of the *Beagle*, closed that gap with his theory of evolution by natural selection.

Except for scriptural literalists, like the young-Earth creationists we learned about in chapter 2, most theologians today admit at least some limited role for natural selection as an agent for biological change over time. What many, but not all, find impossible to accept is the notion that humanity is the product solely of chance and impersonal natural processes. As we saw earlier, a new class of creationists has accepted as valid both the modern cosmological estimates of the age of the universe and much of the evolutionary theory of biology. However, they still insist that the theory of evolution contains gaps which cannot, even in principle, be closed without including outside intervention that they term intelligent design. While these new creationists insist that they are not promoting religion, no objective observer can doubt that explicit vindication of the belief in the traditional, supernatural Judeo-Christian-Islamic creator is their ultimate objective.

Parallel campaigns to incorporate religious concepts into science are now underway on the fringes of physics and cosmology. As we found in the case of Hugh Ross, some contemporary science-theists are familiar with modern physical theories and view them as supportive of their belief in a purposeful creation. They still seek the God of the gaps in places where the theories show possible flaws or where natural explanations are still lacking. As in the case of biology, these science-theists think they may have even found fundamental openings in physics and cosmology that cannot, in principle, be closed by impersonal, material forces alone no matter how much further science progresses.

Those who do not have a good understanding of the scientific method often fail to appreciate that serious theories, even well-established ones, are always growing—evolving—as better data come in. Early calculations meant to describe initial, crude data usually make simplifying assumptions. These then get modified as the new data enable greater detail to be explored. Truly fundamental theories, such as Newton's laws, contain room for these modifications. Indeed, they imply them. The successful modification of a theory enhances its credibility by providing testimony that it has some connection with reality. Many critics of science cannot comprehend this because they have not themselves participated the process of science and seen how it works. "Look at how science keeps changing," they say, as if this were a strike against it.

This illustrates one of the grave problems we all have with scientific understanding today. None of us can follow the details of every theory to the point where we can say we understand it thoroughly. For example, many physicists, myself included, are hopelessly overwhelmed by the mathematics of M-theory. We have no choice but to rely on authority. This is an ironic twist of history when we consider that the modern scientific revolution began four hundred years ago with the rejection of Church authority by Copernicus and Galileo.

Nevertheless, greater understanding can still be obtained, with sufficient effort, when the number of theories we strive to grasp is not so great as to overwhelm and when trusted authorities employing time-tested methods can provide support where we require it. This is the situation at the science-religion interface, where only a few scientific disciplines like physics, cosmology, and biology enter in an important way. A theist looking for gaps in cosmology in which to seek God can and should learn a significant amount of cosmology. Many have done just that, and their claims

thereby become more credible and attract a wider, more knowledgeable audience.

Scientifically competent theists understand that the arguments
for the existence of God from classical theology rely less on
observed fact and more on scholarly rhetoric. The questions they
raise about the source of order in the universe, as exemplified by
natural laws, and how something can come from nothing, are
more philosophical than scientific. Many scientists stay out of the
dispute because they feel that it is insufficiently empirical and only
the empirical is to be trusted. However, I am not one of those who
think that science has nothing to say about ultimate origins, just
because no one was there to make measurements. I will try to
show that science has a lot to say, even if it is not always directly
subject to the empirical testing that most scientists would prefer.

Current theories in physics and cosmology are already well
established by their success in meeting the challenge of severe
testing against existing data. We have every right to logically
extrapolate those theories into the gaps where empirical data are
currently not available, and may indeed never be. Those extrapolations can turn out to be misdirected, so they should not be
treated as scientifically established facts. At the very least, however, they can serve to develop possible scenarios by which the
gaps in current knowledge might plausibly be filled by natural
explanations, thus refuting any assertions that a supernatural
explanation is required by the data.

NOTES

1. William A. Dembski, *The Design Inference* (Cambridge: Cambridge
University Press, 1998); William A. Dembski, *Intelligent Design: The Bridge
between Science and Theology* (Downers Grove, Ill.: InterVarsity Press, 1999).

2. Dembski, *Intelligent Design*, p. 3 (emphasis in original).

3. Lauren Kern, "In God's Country: William Dembski Thought
Baylor University Would Be the Perfect Place to Investigate a Scientific
Alternative to Darwinism. Little Did He Realize He Would Be Crucified
for His Cause," *Houston Press*, December 14, 2000.

4. The External Review Committee Report Baylor University [online], pr.baylor.edu/pdf/001017polanyi.pdf.

5. Kern, "In God's Country"; Eugenie C. Scott, "Baylor's Polanyi Center in Turmoil," *Reviews of the National Center for Science Education* 20, no. 4 (2000): 9–11.

6. Bruce Gordon, "Intelligent Design Movement Struggles with Identity Crisis," *Research News & Opportunities in Science and Theology* 2, no. 1 (2001): 9.

7. Brandon Fitelson, Christopher Stephens, and Elliott Sober, "How Not to Detect Design—Critical Notice: William A. Dembski, the Design Inference," *Philosophy of Science* 66, no. 3 (1999): 472–88; Taner Edis, "Darwin in Mind: 'Intelligent Design' Meets Artificial Intelligence." *Skeptical Inquirer* 25, no. 2 (2001): 35–39; David Roche, "A Bit Confused: Creationism and Information Theory," *Skeptical Inquirer* 25, no. 2 (2001): 40–42; Mark Perakh, "A Consistent Inconsistency" [online], http://www.talkreason.org/articles/dembski.cfm#info [November 2001]; A complete set of links to articles and books on both sides of the issue can be found at the Web site Critical Thought and Religious Liberty [online], http://www/freethought-web.org/ctrl/intelligent-design.html.

8. Dembski, *Intelligent Design*, p. 106.

9. Ibid., p. 156; If $x = \log_2 y$, then $y = 2^x$. The more familiar logarithms have base 10, although the natural logarithm that arises from the exponential function $y = e^x$, where $x = \log_e y$ and e is the irrational number 2.7182818 . . . is more common in mathematics and physics.

10. Claude Shannon and Warren Weaver, *The Mathematical Theory of Communication* (Urbana: University of Illinois Press, 1949).

11. I am grateful to Richard Wein for clarifying for me what Dembski has done.

12. Richard C. Tolman, *The Principles of Statistical Mechanics* (London: Lowe & Brydone, 1938). Later printings available from Oxford University Press.

13. The derivation of the H-theorem, for both classical and quantum systems, can be found in Tolman, *The Principles of Statistical Mechanics*. For an interesting philosophical discussion of the H-theorem as it relates to the arrow of time, see Huw Price, *Time's Arrow and Archimedes' Point: New Directions for the Physics of Time* (Oxford: Oxford University Press, 1996). I also discuss this connection in Victor J. Stenger, *Timeless Reality: Symmetry, Simplicity, and Multiple Universes* (Amherst, N.Y.: Prometheus Books, 2000).

14. Dembski, *Intelligent Design*, p. 168.

15. Peter B. Medawar, *The Limits of Science* (New York: Harper & Row, 1984), p. 79; Thanks to Lloyd Davidson for supplying me with this reference.

16. Kern, "In God's Country."

17. Dembski, *Intelligent Design*, p. 170.

18. Ibid., p. 166.

19. Emile Borel, *Probabilities and Life*, trans. M. Baudin (New York: Dover, 1962), p. 28.

20. Dembski, *Intelligent Design*, p. 178.

21. Howard J. Van Till, "E. Coli at the *No Free Lunchroom*" [online], http://www.aaas.org/spp/dser/evolution/perspectives/vantillecoli.pdf.

22. Ibid.

23. William Dembski, *No Free Lunch: Why Specified Complexity Cannot Be Purchased without Intelligence* (Lanham, Md.: Rowman & Littlefield, 2002).

24. David H. Wolpert and William G. Macready, "No Free Lunch Theorems for Optimization," *IEEE Transactions on Evolutionary Computation* 1 (1997): 67–82; Joseph Culberson, "On the Futility of Blind Search: An Algorithmic View of 'No Free Lunch,'" *Evolutionary Computation* 6 (1998): 109–27.

25. Richard Wein, "Not a Free Lunch But a Box of Chocolates" [online], http://www.talkorigins.org/design/faqs/nfl/#summary.

26. H. Allen Orr, review of *No Free Lunch*, by William Dembski, *Boston Review*, summer 2002.

27. Jason Rodenhouse, "Probability, Optimization, and Evolution," *Evolution* 56, no. 8 (2002): 1721–22 [online], http://www.math.ksu.edu/~jasonr/dembski.htm.

28. Jeffrey Shallit, Review of *No Free Lunch*, by William Dembski, *Biosystems* 66, no. 1–2 (2002): 93–99 [online], http://www.math.uwaterloo.ca/~shallit/bookrev.html.

29. Michael J. Behe, *Darwin's Black Box: The Biochemical Challenge to Evolution* (New York: Free Press, 1996), p. 42.

30. Ibid., p. 45.

31. Ibid., p. 39.

32. H. J. Muller, "Reversibility in Evolution Considered from the Standpoint of Genetics," *Biological Reviews* 14 (1939): 261–80.

33. Robert Dorit, review of *Darwin's Black Box*, by Michael Behe, *American Scientist* (September–October 1997); H. Allen Orr, "Darwin v. Intelli-

gent Design (Again): The Latest Attack on Evolution Is Cleverly Argued, Biologically Informed—and Wrong," *Boston Review* (1998); Niall Shanks and Karl H. Joplin, "Redundant Complexity: A Critical Analysis of Intelligent Design in Biochemistry," *Philosophy of Science* 66 (1999): 268–98; Kenneth R. Miller, *Finding Darwin's God: A Scientist's Search for a Common Ground between God and Evolution* (New York: HarperCollins, 1999).

34. Behe, *Darwin's Black Box*, p. 185.

35. Miller, *Finding Darwin's God*, p. 143.

36. Roger Penrose, *The Emperor's New Mind: Concerning Computers, Minds, and the Laws of Physics* (New York: Oxford University Press, 1989); Roger Penrose, *Shadows of the Mind: A Search for the Missing Science of Consciousness* (Oxford: Oxford University Press, 1994).

37. Victor J. Stenger, *The Unconscious Quantum: Metaphysics in Modern Physics and Cosmology* (Amherst, N.Y.: Prometheus Books, 1995), chap. 10.

38. See Stuart R. Hameroff, *Ultimate Computing: Biomolecular Consciousness and Nano-Technology.* (Amsterdam: North-Holland, 1987). Hameroff proposes that large-scale quantum effects might occur in the microtubules attached to cells. These have an inner diameter of typically 14×10^{-9} meter, even larger than the synaptic gap. Penrose in *Shadows of the Mind* has enthusiastically backed this notion, however it has not yet received convincing empirical support. They do not explain why the microtubules attached to brains cells are any different from those of the bladder or big toe.

39. Penrose, *Shadows of the Mind*, p. 169.

40. Taner Edis, "How Gödel's Theorem Supports the Possibility of Machine Intelligence, " *Minds and Machines* 8 (1998): 251.

41. By totally random I mean not pseudorandom. A pseudorandom number is generated by an algorithm, and so the robber can, in principle, predict its value if he knows the algorithm. However, assuming, as is conventional, that quantum mechanical fluctuations are unpredictable, a quantum source can be used as the random number generator.

42. Roche, "A Bit Confused."

43. Edis, "Darwin in Mind."

44. Roche, "A Bit Confused," quoting Darwin, *The Origin of Species*.

45. Mark Perakh, "A Presentation without Arguments: Dembski Disappoints," *Skeptical Inquirer* 26, no. 6 (2002): 31–34.

FIVE

THE GOD OF
FALLING BODIES

Reality is that which, when you stop believing in it, doesn't go away.
—Philip K. Dick, 1972

GALILEO, NEWTON, BENTLEY, AND LEIBNIZ
CHAT ON THE INTERNET

In order to explore some of the thinking processes involved in the current dialogue between science and religion, I have imagined the following parable. The characters in my parable are modern-day versions of Galileo, Isaac Newton, and Gottfried Wilhelm Leibniz. Also included is a lesser-known historical figure, theologian Richard Bentley, with whom Newton corresponded.

Galileo is pictured as a modern-day experimental physicist, performing increasingly precise experiments with falling bodies at the Leaning Tower of Pisa. I imagine him rapidly communicating his results by e-mail to Newton in Cambridge, who is contemporaneously developing his laws of motion and gravity. Of course, Galileo preceded the other characters by two generations, so this interchange is obviously not historical. Furthermore, although both men were brilliant theorists and experimentalists, I am going to impose a modern division of labor and have Galileo be strictly an experimentalist and Newton a theorist. Galileo will have the best modern equipment at his disposal, and I will imagine each thinking like a scientist of today, not one of the sixteenth and seventeenth centuries.

The scene opens with Bentley discussing Galileo's experimental results and their theological implications with Newton and Galileo over the Internet. I show Newton doing his best to explain the data by means of natural laws and, like typical modern theoretical physicists (that is, those who do not attempt to write popular books), not fretting too much about theology. While the historical Newton branched off into theology and alchemy, this was not until later in his life. In my scenario he is still a typical (for today) young researcher, impatient with philosophizing and eager to get on with his work with minimum distraction.

Bentley seeks the God of the gaps, looking for places where Newton's theories seem to leave room for the creator to impose his will. He exhibits the general misunderstanding and resulting distrust of science that typifies the modern intellectual, who is intelligent but lacks scientific training and, worse, has little comprehension of scientific method. In this story, Galileo expresses religious skepticism more openly than he could in his day, but he would have no trouble getting away with such skepticism today. Finally, Leibniz joins the discussion near the end. He represents the new crop of science-theists who, unlike Bentley, know their science and mathematics but still think they see God's hand in physics and cosmology. Darwin also pokes his head in at the end.

In his initial, crude experiments, Galileo measures the times, t, that it takes cannonballs of various masses to drop from balconies in the tower at different heights h. He makes a graph of h versus t

and shows that the data fit a parabolic curve, $h = kt^2$, with k a constant equal to 4.9 when h is measured in meters and t in seconds, independent of the mass of the cannonball.

When Newton sees these results, he e-mails Bentley and Galileo:

"Dear Friends. This is exactly what is predicted by my laws of motion and gravity. My second law of motion is $F = ma$, where F is the force on a body, m is its mass, and a is its acceleration. Putting it together with my law of gravity gives $a = g$, where $g = 9.8$ meters per second squared is the acceleration due to gravity, independent of the mass, m. Using the methods of calculus, which I invented, despite the claims of that upstart Leibniz, I then get $h = kt^2$ where $k = g/2 = 4.9$."

Bentley finds Newton's explanation difficult to understand: "Isaac, as usual I am clueless as to what you are talking about. Why should this 'calculus' of yours, with all its magical symbols, have anything to do with reality?"

Newton responds, "Richard, I don't know why, but it seems to. I frame no hypotheses. I just calculate and compare my calculations with the data."

In the meantime, Galileo continues his experiments with objects other than cannonballs and discovers something new. He tries a crumpled-up piece of paper, along with a rock. Releasing them at the same time from the same height, the paper hits the ground after the rock. A sheet of paper and a feather take even longer.

When Bentley sees this result, he excitedly types: "See, Isaac, your theory is incomplete. God is acting to hold up the paper and the feather. This explains how birds and angels fly! God wills it."

Galileo, who has been quiet so far except for supplying the data, butts in: "I've never seen any angels, even with my telescope. But birds must fly by taking advantage of the upward force of the air, as da Vinci has suggested."

Newton does not take long to respond: "In the original experiments, Galileo was dropping heavy objects—cannonballs. So I imposed a simplifying assumption about air resistance, which I guessed would be small in that case. In general, however, the air is expected to exert an upward force that subtracts from the downward force of gravity, and this will be important for lighter objects. This resistive force depends on the velocity at which the body falls. I have modeled it as proportional to the square of the velocity and determined an air resistance coefficient from the data, which varies from object to object. Using some more calculus, I have obtained a fairly reasonable fit to Galileo's data, as you can see by the attached graphs."

Bentley is not too impressed: "That looks like a pretty ad hoc procedure to me. And so complicated! Only two people in the world can make such a calculation, Newton and Leibniz. Are we to rely on the authority of just the two of you? I prefer to rely on the authority of scriptures and the Church fathers. They provide a much simpler explanation that even the humblest peasant can understand, namely that God directs the motion of all things, from falling leaves to flying birds."

Galileo is a bit annoyed: "I think I can manage this calculus, too. After all, I am a professor of mathematics! But, more important, where in the God theory can you obtain the detailed, quantitative results Isaac has here? He can make all kinds of predictions about falling bodies and projectiles. Even the biblical prophets could not do that."

"They were men of peace, not interested in bombs and cannonballs," Bentley reverently but irrelevantly replies.

Things only get worse when Galileo reports another strange anomaly. His falling bodies do not hit the ground at a point directly below the release point, as marked precisely by a plumb bob, but slightly off to the east. He is careful to show that this is not an effect of winds.

"Aha," Bentley cries, when the data appear on the Web. "Here is incontrovertible evidence for God's action. The creator is blowing the objects off to the east."

"Why would he do that?" Galileo questions.

"This is just one of those mysteries we were not meant to understand," Bentley answers.

Newton scratches his head but soon realizes what is happening.

"My previous calculations assumed that Earth is not accelerating. In fact, the rotation about its axis constitutes an acceleration. When I properly add this term to my equations, I get exactly what Galileo observes," Newton types.

"More ad hoc fixes and fudge factors," Bentley retorts scathingly. "And look at those equations now. How complicated can they get? What purpose are you serving with all these esoteric symbols. It looks to me like you are starting a new cult, and you know what the Church thinks of cults!"

"Christianity was once a cult," Galileo comments sourly.

Newton tries to cool things off. "Bentley, it is too bad you have not been able to follow my mathematics. (Damn these lousy schools.) If you could do the maths, you would see that my equations already contained the solutions to all the problems raised by Galileo's increasingly more precise measurements. The F in $F = ma$ represents the sum of the forces on a body. The term I added for air resistance in retrospect should have been included all along. Similarly, the a in $F = ma$ must include the acceleration of Earth. Putting in the correct acceleration, we again get what Galileo measures. What happens in the present case is that the body at its point of release has a greater eastward component of velocity than a point on the ground, so it drifts to the east. And here is a prediction! If Galileo does experiments with cannon balls shot straight up in the air, they will drift to the west."

Back in Italy, Galileo is presented with a huge grant from Cosimo de Medici, from which he purchases lasers and a highly accurate atomic clock. Repeating his experiments, he finds that, even after corrections for air resistance and Earth's rotation, the g in Newton's equations is not a constant but depends on the height of the tower balcony from which objects are dropped.

Once again, Bentley goads Newton: "This surely proves that your theory is, at best, an approximation and so cannot be related in any important way to 'ultimate reality.' Each time our friend Galileo makes a better experiment, you have to modify your equations to make them agree with his data. What are you going to do now about this nonconstant value of g?"

"Well, if you could follow the maths, you would see that this, too, is in my equations. When I conceived the law of gravity, I realized it applies to objects far from Earth, such as the Moon, as well as apples and leaves falling from trees. The Moon, in a sense, is falling toward Earth like an apple; but, because of its speed in orbit, it falls around Earth without ever hitting it. From estimates of the Moon's distance and the time it takes to go around Earth, one month, I was able to infer that the force of gravity, and thus the acceleration of a falling body, will decrease as the square of its distance to the center of Earth. In fact, my law of gravity reads $F = GmM/r^2$ as the force between two bodies of masses m and M whose centers of gravity are separated by a distance, r, where G is a constant determined from the data.

"The resulting acceleration," Newton patiently points out, on a body of mass m toward Earth is then $g = GM/r^2$ where M is the mass of Earth and r is the distance to the center. The variation with r is normally unmeasurable near Earth's surface, since the difference between r and the radius of Earth, R, is small. So we are justified in neglecting it for most practical purposes. However, Galileo was able to detect the variation with his lasers and atomic clock."

"If I can't read your equations, neither can the great majority of the human race," Bentley responds. "How are you ever going to convince them?"

Newton sighs. "OK, let me try to explain the significance of what I have done in words, which are unfortunately less precise than the maths. I have provided techniques that enable a sufficiently trained person to make quantitative calculations of accurate measurements that agree with all the data. These equations also enable that person to make predictions about the motions of bodies that can be later tested by experiments. I hope Galileo and others will carry out these tests of my theories. My good friend Edmund Halley has just informed me that my equations predict that the recently observed comet will return again in seventy-five years. Unfortunately, we will not be here to see if this prediction comes true. Even if some of these predictions fail, this could simply mean that I have once again made too many simplifying assumptions in my calculation, as I did when originally neglecting air friction or Earth's rotation. The comet prediction should be an accurate one, however, since neither Halley nor I can think of any factors that may mess it up."

Newton then comes to a sudden realization. "Bentley, you have continually derided the fact that I did not anticipate some of Galileo's measurements before they were made. But rather than taking this as a point against the validity of my theories, you should regard it as a point for!"

Bentley blinks. "Come again?"

"The fact that even I, the inventor of the theories, did not realize all their implications indicates, rather strongly I think, that they indeed have something to do with reality. In fact, you might say that I was not the inventor of the theories at all, but rather their discoverer. They were out there in nature waiting for someone brilliant like me to come along to find them. Let me contrast my theories with yours, dear Bentley, that God has done it all. You claim your theory is simpler, and so more preferable, more likely to be correct than my complicated calculus equations."

"But is it simpler?" Newton asks. "I have been able to classify a wide range of phenomena, on Earth and in the heavens, with a few assumptions that are very simple in their own right. The complications you worry about are only in the manipulations, which admittedly require some inborn talent comparable to playing a musical instrument well."

Galileo then jumps in with a thought: "Perhaps, someday, humans will possess machines that will do these calculations for them. Then all they will have to do is put in the initial positions and velocities, and predict the future motion of all bodies. If Lucretius is correct—that everything is made up of atoms—then everything will be predictable."

Suddenly Bentley breaks out into a broad grin and excitedly types: "Even if you are correct, and everything that happens in the universe can ultimately be predicted by some huge machine, the hand of the creator was still involved. You have just written down some esoteric equations, but you have not told me where those equations come from. I think it is all pretty obvious. They came from God!"

"Why did they have to come from anything?" Galileo interjects.

"Everything comes from something."

"And God, where did he come from?"

"Well, God is the exception. As Aristotle said, the First Cause, uncaused."

"Why can't that exception be the universe itself?"

Bentley does not answer, since he has become troubled by another thought: "I don't think I like this idea after all. What happens to free will?"

"I will leave it to you theologians to figure that one out," Galileo responds.

Newton has not said much for a while and now speaks up: "Actually, now that you have distracted me from my research and dragged me into a theological discussion, I must admit that my theory does not account for everything. Remember I said that my law of gravity does not give the value of G. I have to get that from the data. Also, recall that the Moon is like a falling object. My equations will tell you that the Moon's orbit around Earth is elliptical, but they do not give the orientation of the axes of the ellipse."

"Ah, better yet!" Bentley exclaims. "We are back exactly to the God of the Bible. He creates the universe with its matter and light. He commands this matter and light to obey certain natural laws, which you scientists are now beginning to discover. But the creator sees to it that the laws do not preordain all that happens. Humans then have the free will to act, from which we get evil despite God's innate goodness. All this freedom, however, can lead to things getting out of hand. So God acts whenever necessary to keep the universe and mankind moving on track toward the ultimate realization of his divine plan."

Just then an e-mail comes in from Leibniz in Germany: "I just happened to get wind of this discussion while surfing the Web. I have looked at Newton's equations on his Web page and can confirm that they fit Galileo's data. In fact, I did invent calculus independently and used my own methods, which I think are superior, especially in terms of notation."

Newton: "Balderdash!"

"In any case," Leibniz continues, "I have to go along with Bentley that God's purpose is evident in all that is being uncovered here. Let's take Newton's constant, G, in his theory of gravity. He admits that his theory does not give its value, that it must be discovered by experiment. I am sure that Bentley will agree that it must be set by God."

Bentley responds, "Indubitably."

"But I have more," Leibniz types. "I think I can prove that God has set this value of G very precisely for the divine purpose of making human life possible. Newton's equations, which I truly do admire despite their lamentable notation, have allowed me to calculate, with my own better methods, the effect that different values of G would have on the orbit of Earth. Earth might have been farther from the Sun and too cold for life, or closer to the Sun and too hot."

Newton replies, "Yes, yes. This is just the $r^3/T^2 = a$ constant law discovered observationally by Kepler, which I have already proven from my theory. Note that if G were different, we could have the same orbital radius r, as now with just a different orbital period, T."

"I agree," says Leibniz. "But with all the values of r and G to chose from, how unlikely it is that a random selection would have given just the right values we need for our existence? The universe seems to be fine-tuned for humanity. Suppose we had a world in which a year was not 365 days. I shudder to think of what this would do to the seasons. I would wager that human life would again be impossible. As far as I can see, only the exact value of G, and the specific values of r and T we have, would allow for human life. God has obviously chosen these numbers carefully and created this as the best of all possible worlds."

"Or, the worst," Galileo replies. "It could all be just one big accident."

Just then a message comes in from someone who has been listening in on this discussion ("lurking") without commenting, Charles Darwin: "The universe is not fine-tuned to humanity. Humanity is fine-tuned to the universe."

SIX

THE UNCREATED
UNIVERSE

Who are we? We find that we inhabit an insignificant planet of a hum-drum star lost in a galaxy tucked away in some forgotten corner of a uni-verse in which there are far more galaxies than people.

—Carl Sagan, 1983

THE FIRST TEN MICROSECONDS

In an article in the *New York Times* on January 28, 2001, science reporter James Glanz wrote:

Ask a philosopher, a theologian, an artist, or a composer how close humanity is to understanding the mystery of cosmic creation, and you are liable to get an answer that is majestic, inspiring, and extremely imprecise. Ask a physicist the same question and the answer will be much more cut-and-dried: about 10 millionths of a second.[1]

While he was not responding directly to Glanz, Richard Dawkins has tried to counter the common impression, to which Glanz may have further contributed, that science has removed all the beauty and mystery from life:

> The feeling of awed wonder that science can give us is one of the highest experiences of which the human psyche is capable. It is a deep aesthetic passion to rank with the finest that music and poetry can deliver. It is truly one of the things that make life worth living and it does so, if anything, more effectively if it convinces us that the time we have for living is quite finite.[2]

I do not think science has to make any apologies. It looks at the world and tells it like it is. And we all live longer, better lives because of this dispassionate view. Sure, it commands awe and provides inspiration. Still, I would rather be operated on by a surgeon who sees me as an assemblage of atoms than one who lovingly tries to manipulate what he or she imagines are my vital energy fields. Dawkins himself has been particularly eloquent in getting across the message that science does not paint a picture of a universe that always fulfills human wishes. Indeed, it paints a more wondrous sight that goes far beyond human fantasies and petty concerns.

Glanz was reporting on a new experiment about to commence operation at the Brookhaven National Laboratory on Long Island in New York. Gold nuclei moving near the speed of light were to be smashed against other gold nuclei, producing the enormous energy densities that, according to current cosmological theory, existed during a few microseconds after our universe began. This is made possible by the high electric charge of the gold nucleus and the enormous electric field that develops when two such nuclei come into near contact. Under these conditions, matter exists in a state called the *quark-gluon* plasma. In a more familiar plasma,[3] such as exists in the ionosphere of Earth's atmosphere, electrons are ripped from atoms leaving behind electrically charged nuclei and electrons. In the Brookhaven quark-gluon plasma, not only are the original gold atoms ripped apart, but so are their nuclei and the protons and neutrons inside these nuclei. What remains is a highly dense mixture of *quarks*—the constituents of nucleons—and *gluons*, the particles responsible for the force between quarks. Under normal conditions, quarks remain bound inside protons and neutrons. However, for a small fraction of a second, the Brookhaven experiment is designed to produce matter as it existed in the first ten microseconds.

The theory that describes the quark-gluon plasma is called *quantum chromodynamics* (QCD) and is part of the wider *standard model* of elementary particles and forces developed in the 1970s. This model proposes that a small number of elementary particles (quarks and *leptons*) comprise the basic units of matter while the forces between these particles result from the exchange of other elementary particles (*bosons*).[4] The Brookhaven experimenters will fill in further details and look for anomalies, but so far QCD and the standard model have passed every experimental test. The current expectation is that no new physics beyond the standard model will be discovered until the Large Hadron Collider (LHC) now being built in Geneva, Switzerland, at the European Organization for Nuclear Research (CERN) begins taking data around 2006. Of course, physicists will keep looking for new physics with existing technology.

While science continually uncovers new mysteries, it has removed much of what was once regarded as deeply mysterious. Although we certainly do not know the exact nature of every component of the universe, the basic principles of physics seem to apply out to the farthest horizon visible to us today.

With powerful telescopes and other instruments, we can observe and study galaxies billions of light-years distant. These galaxies look different from ours in several ways; for example, they have many more young stars. Even as I write this, the media are reporting the first results from the Chandra X-ray space telescope, which indicate far more high-energy radiation from distant galaxies, viewed as they were billions of years ago, than those closer by and "only" a million years or so from our own epoch. The data offer strong support for a universe that has evolved over time, one that is not the "firmament" implied by the Bible. Not only has life evolved, so has the universe itself.

However, while early and distant galaxies are structurally different from the ones nearer home, they exhibit the same basic physical processes. For example, the spectral lines of atomic hydrogen emitted by other galaxies, near and far, are identical to those observed in the Sun and in laboratories on Earth. They are just shifted to longer wavelengths or lower frequencies (the "red shift") by the Doppler effect that results from the high speeds at which the galaxies move away from us. (The Doppler shift is the familiar experience in which the sound of an ambulance siren is higher pitched as the ambulance rushes toward us, and then becomes lower pitched as it moves away). Similarly, the faster a galaxy is receding from us, the farther away it gets as time goes on. Thus, the most distant galaxies are those that are receding at the highest speeds. Another interesting recent discovery by the Chandra X-ray telescope provides evidence that giant black holes are likely sources of the distant X-rays that it has detected. These deep-space X-rays were predicted to exist by the same equations from Einstein's theory of general relativity that explain minute gravitational effects observed in our own solar system much closer to home.

In other words, while the detailed structure of the universe has evolved over billions of years, the basic laws of physics have not. They have apparently remained in force as far back in time as we have been able, so far, to peer. However, while we cannot look directly into the heart of the early universe, our existing knowledge can be used to infer the physical processes that took place within a tiny fraction of a second after the start of things—over ten billion years ago. This extrapolation beyond the observable is not as unreliable a procedure as you might think. While any such extrapolation can turn out to be wrong, science has a successful track record in applying its established theories to new situations.

For example, we were able to send men to the Moon and get them back by assuming that the same laws of physics applied on the Moon as on Earth. Indeed, this may have been Newton's greatest insight—that his laws of motion and gravity apply for the Moon falling around Earth as much as for the apple falling off a tree. Theories of the early universe have been precisely tested in laboratory experiments that, like Brookhaven, mimic conditions at that time. We have no reason to think these theories should not apply, although we must make whatever observations and checks that are possible.

The standard model, when combined with equally conventional cosmological models, provides a picture of the early universe that is consistent with the very detailed measurements made in recent years on the cosmic microwave background. This radiation has long stopped interacting significantly with the other matter of the universe, cooling off (considerably) over billions of years as a result of the universe's expansion but remaining otherwise pristine in its structure. Observations of that structure have beautifully corroborated the cosmological picture developed over the last twenty years.

Evidence for the expansion of the universe was first found by Edwin Hubble in the 1920s with his observations of the red shift of galaxies. In fact, it was Hubble who first recognized that many (but not all) of the fuzzy "nebulae" he and other astronomers saw in the sky were not part of the Milky Way but were separate galaxies of stars in their own right. The detection of the cosmic microwave background by Arno Penzias and Robert Wilson in 1965 provided the first confirmation for the *big bang* model in which the universe is seen as the expanding remnant of an explosion of matter and energy that occurred 13 billion years ago.

A number of theoretical difficulties with the big bang led to the development, in the 1980s, of a supplementary model called *inflation*.[5] This model proposed that the currently observed, almost (but not quite) linear expansion was preceded by an exponential expansion during which the universe grew by many orders of magnitude in a time interval many orders of magnitude less than a second.

Although, as creationists will tell you, no physicist was present billions of years ago to observe inflation, the quantitative theory had measurable implications that could be tested. These tests were sufficiently stringent to have been capable of falsifying the inflationary theory of the early universe. As we will see, the theory of inflation has so far passed every test.[6]

Early observations indicated that the cosmic microwave background radiation was highly uniform, with the same temperature of 2.7 degrees Kelvin being measured in all directions. This uniformity was a puzzle and provided one of the original motivations for proposing inflation. In the linear big bang scenario, different parts of the sky would never have been in contact and thus unable to reach the equilibrium indicated by these regions having the same temperature. With exponential inflation, those regions were easily in original contact.

Still, while high uniformity is good for inflationary theory, too much uniformity is bad. Calculations with the inflationary model indicated that the observed temperature in different directions in space had to vary by about one part in one hundred thousand or else galaxies would never have formed. A greater observed uniformity would have implied that the matter to build galaxies was far too smoothly distributed to accumulate by gravity into localized objects, such as galaxy clusters, in the time since inflation ended.

As we learned in chapter 3, the precise quantitative variation from uniformity that inflationary theory required was verified in 1992 by the Cosmic Microwave Background Explorer (COBE) satellite. Recall that Christian physicist Hugh Ross called this the "discovery of the century" since it confirmed that the universe had a beginning and thus, in his mind, was created by a supernatural being.

Observations since 1992 have further substantiated the validity of inflationary big bang cosmology. At this writing, a remarkable new satellite observatory called the Microwave Anisotropy Probe (MAP) has just been launched that will measure the temperature to twenty millionths of a degree Kelvin in angular intervals of 0.23 degree on the celestial sphere.[7] This should lead to an even deeper understanding of the early moments of the big bang. Many of the details of the early universe, and questions such as the nature of the *dark matter* and *dark energy*, are very sensitive to the structure of the microwave background.

WAS THE CREATION OF THE UNIVERSE A MIRACLE?

People have a hard time imagining how the universe can possibly have come about by anything other than a miracle, a violation of natural law. The intuition being expressed here is at least twofold. First, it is widely believed that something cannot come from nothing, where that "something" refers to the substance of the universe—its matter and energy—and "nothing" can be interpreted in this context as a state of zero energy and mass. Second, it is also widely believed that the way in which the substance of the universe seems to be structured in an orderly fashion, rather than simply being randomly distributed, could not have happened except by design.

By the way, the universe is not as orderly as most people think. We live on a small pocket of order, Earth, and we see stars and galaxies in the sky that exhibit what seems to be a lot of structure. However, as we will see in more detail below, the visible matter of the universe is only about 0.5 percent of all the matter in the universe. Much of the rest, as best as we can tell, has little more structure than the cosmic microwave background, which we recall is smooth to one part in a hundred thousand.

Let us look at the physics questions implied by common intuition. If we hypothesize that the universe is an isolated or "closed" system, meaning nothing going in and nothing coming out, then both the first and second laws of thermodynamics would seem to have been violated when the universe, as we know it, came into existence. The first law is equivalent to matter-energy conservation, and a reasonable question is: Where did the current matter and energy of the universe come from?

As best as we can tell from current observational data, the total kinetic energy of motion exactly balanced by the negative energy of gravity. As Stephen Hawking explains it:

There are something like ten million million million million million million million million million million million million million million (1 with eighty zeroes after it) particles in the region of the universe that we can observe. Where did they all come from? The answer is that, in quantum theory, particles can be created out of energy in the form of particle/antiparticle pairs. But that just raises the question of where the energy came from. The answer is that the total energy of the universe is exactly zero. The matter in the universe is made out of positive energy. However, the matter is all attracting itself by gravity. Two pieces of matter that are close to each other have less energy than the same two pieces a long way apart, because you have to expend energy to separate them against the gravitational force that is pulling them together. Thus, in a sense, the gravitational field has negative energy. In the case of a universe that is approximately uniform in space, one can show that this negative gravitational energy exactly cancels the positive energy represented by the matter. So the total energy of the universe is zero.[8]

Actually, Hawking is referring only to the total kinetic and potential energies. While these cancel, energy is still carried in the mass of particles. This is the rest energy given by Einstein's equation $E = mc^2$. However, in the inflationary scenario, the mass-energy of matter was produced during that rapid initial inflation. The field responsible for inflation has negative pressure, allowing the universe to do work on itself as it expands. This is allowed by the first law of thermodynamics. Some small amount of energy was required to trigger inflation, but was a fluctuation allowed by quantum mechanics.

In other words, *no* energy was required to "create" the universe. The zero total energy of the universe is an observational fact, within measurement uncertainties, of course. What is more, this is also a prediction of inflationary cosmology, which we have seen has now been strongly supported by observations. Thus we can safely say,

No violation of energy conservation occurred if the universe grew out of an initial void of zero energy.

Another common belief is that the formation of order by natural processes is impossible. This is the old argument from design, which was discussed in detail in chapter 3. Here it appears as the intuitive claim that the second law of thermodynamics requires that the universe begin in a state of low entropy (high order) and evolve toward a final state of ultimately maximum entropy (low order)—the so-called heat death of the universe. Creationists have asserted that even if local order can occur naturally, supernatural design is evident in the existence of the highest level of order, that is, lowest entropy, at the "creation."

This argument had great weight in the nineteenth century, when the universe was assumed to be the biblical firmament of fixed stars. However, we now know that the universe is expanding. As shown in appendix C, treating the universe as a sphere of radius R,[9] the entropy of the universe increases linearly with R. However, the maximum allowable entropy of the universe increases with the square of R. As shown in figure 6.1, this allows increasing room for order to form locally.

This can be easily understood from the following mundane example. Suppose, each day you empty your kitchen waste basket into your yard. Pretty soon the yard will have no room left for trash. So you buy up the surrounding property and start dumping there. As long as you keep that up, expanding your property perimeter, you can always make your house more orderly by simply dumping your rubbish (entropy) to the outside.

To make this more quantitative, and thus more precise, I once again need to get a bit technical. The formation of order on Earth is illustrated in figure 6.2. For each visible photon that Earth receives from the Sun, it emits twenty infrared photons back to the universe. This is simple energy conservation. Earth is in thermal equilibrium with a surface temperature of 300 degrees Kelvin, kept there by energy from the Sun.[10] The Sun is also in thermal equilibrium, with a surface temperature of 6,000 degrees Kelvin, maintained by nuclear processes at its core. Both the Sun and Earth radiate a spectrum of photon energies, but, on average, a photon emitted from the Sun is twenty times as energetic as one from

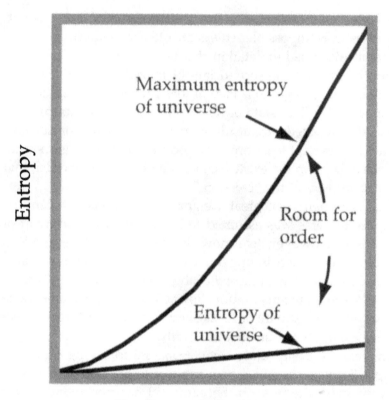

Radius of the Universe

Fig. 6.1. The entropy of the universe as it expands from Planck dimensions increases linearly with radius while the maximum allowable entropy increases quadratically. Thus, although the universe starts in chaos at maximum entropy, there is increasing room order to form locally as the universe expands.

Earth since its surface temperature is twenty times higher than Earth's.

Each photon can be regarded as one bit of entropy, using the units of entropy defined by Shannon (see chapter 4). In this process, the Sun loses one bit of entropy and Earth loses a net of nineteen bits. Thus the Sun becomes more orderly by one bit, Earth more orderly by nineteen bits, and the rest of the universe more disorderly by twenty bits. The local ordering of the Sun and Earth is made possible by the fact that the maximum entropy of the universe (in figure 6.1) is much greater than its actual total entropy, as

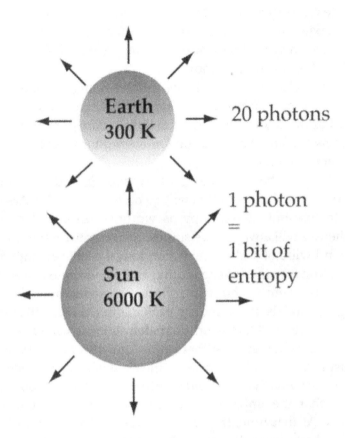

Figure 6.2. The formation of local order in the Earth-Sun system. For each photon that the Sun emits and is absorbed by Earth, Earth returns twenty to the rest of the universe, and its net order increases by nineteen bits.

described above, so the universe has increasing room to gain entropy as it expands.

A simple calculation shows that more than ample room exists for the formation of the order in all the galaxies in the universe. The photons emitted by Earth are absorbed by the Cosmic Microwave Background (CMB), which currently has a temperature of 3 degrees Kelvin but is not in thermal equilibrium, cooling as the universe expands. Since Earth was formed, it has emitted about 10^{54} photons, thus increasing the entropy of the CMB by that amount. However, the CMB contains about 10^{87} photons. If we estimate ten planets for every star, 10^{11} stars for every galaxy, and 10^{11} galaxies in the visible universe, then the ordering of those planets has increased the entropy of the CMB by 10^{76}, or only one part in 10^{11} of its total entropy.[11]

We can also understand this process using the information-theory language discussed in chapter 4, where information was defined as the change in entropy in bits. In Earth-Sun example, the Sun gains one bit of information for every photon lost, while Earth gains nineteen bits.

As long as the universe keeps expanding, and we now have good reason to think that this will go on forever, we always have a place to toss out our entropy as we organize ourselves locally. Whether we will always have sufficient energy to do this is another question I will not address here. For now, we have enough.

The universe is now expanding, and we have no evidence that it ever underwent a contracting phase. We can use our existing cosmological models to extrapolate back in time to when the universe was a sphere 10^{-35} meter in diameter, what is called the *Planck length*. The models suggest that at this time the universe was indistinguishable from a black hole of the same size (see appendix A). Since a black hole has maximum entropy for an object of its size, it follows that the universe had maximum entropy at this early moment. At that time, the universe was as disorderly as it could possibly have been. It was without order—without design. If a creator existed, any information he may have inserted into the universe prior to that time would have been lost. Thus,

No violation of the second law of thermodynamics was required to produce the universe.

In short, no miracle, no violation of any known principles of physics, need have occurred at the creation.[12] In fact, the data are just what would be expected for a universe that came into being without design or cause.

FINE-TUNED FOR YOU AND ME

In chapter 3, I briefly mentioned the *fine-tuning* argument for the existence of purposeful design in the universe. We saw that this is another of the variations that have recently appeared on the ancient, already well refuted argument from design. The so-called anthropic coincidences are claimed as evidence for a universe that was created with humans in mind.

I have covered the fine-tuning claims extensively in two previous books, *The Unconscious Quantum* and *Timeless Reality*. Here, to avoid repetition, I will only summarize the points made there and concentrate on important new developments from cosmology that have provided a tentative answer to one particular "coincidence," which, until now, has been perhaps the most puzzling of all—at least to physicists.

Most of the anthropic coincidences bear on the manufacture of carbon and other heavy elements that occurs in stars and the time needed for life to evolve once these elements are released into space and coalesce into planets. In the entire periodic table of chemical elements, which one can find hanging on the wall of any chemistry classroom, only the first three lightest elements—hydrogen, helium, and lithium—were produced in the early universe. All the elements heavier than lithium, which are called "heavy elements" in an astronomical context, were manufactured in massive stars.

Over the billions of years of the lifetime of a star, hydrogen nuclear fusion provides its main source of energy. Once its hydrogen is used up, a star collapses and its internal pressure rapidly increases. A less massive star, like our Sun, eventually reaches a stable state called a *white dwarf*. A more massive star continues to collapse, and heavy elements are fabricated by nuclear processes in the intense heat that develops. At some point the immense pressure builds up to the point where the star explodes as a *supernova*, spraying its matter into interstellar space and leaving a *neutron star* behind. Planets composed of heavy elements can then assemble from this matter as it is drawn together by gravity.

It has long been known that the production of carbon in stars depends sensitively on certain parameters in nuclear reactions. If those parameters had been slightly different, sufficient carbon may not have been available for life as we know it.

Obviously, the parameters were such as to produce sufficient carbon for life as we know it; however, is carbon necessary for every conceivable from of life? Hugh Ross seems to think so:

> If you want physicists (or any other life-forms), you must have carbon. Boron and silicon are the only other elements on which complex molecules can be based, but boron is extremely rare, and silicon can hold together no more than about a hundred amino acids. Given the constraints of physics and chemistry, we can reasonably assume that life must be carbon based.[13]

But, conceivably, boron would not be so rare with some other set of parameters determining the production of elements in the nuclear reactions inside of stars. And, why must life under every circumstance be based on amino acid chemistry? Ross and other theists seem to be blind to the possibility of forms of life other than those based on carbon, perhaps because of the religious doctrine that we were all made in the image of God.

From what we now know, "life" is the label we assign to a material structure that exhibits a certain set of qualities and characteristics when those structures have reached a high level of complexity. With the physical laws and constants of our universe, heavy-element chemistry—not necessarily carbon-based—may be the only available platform for life, although we can't be sure with only one form of life to study. In opening the possibility of alternative universes with different laws and constants, we can hardly even speculate on what other forms life might take. And it is pure speculation to suggest that no form of life other than our own is possible under all circumstances.

About the best we can do with existing knowledge is consider what the universe might be like if it had the same basic physics equations but with different values of the "constants" that go into those equations. We can use those same equations to calculate various properties that the universe might have under those conditions.

If we limit ourselves to life based on chemistry, then one obvious property that a universe with life must possess is a long lifetime for stars to allow life to evolve from whatever elements may be present in the interstellar medium. In appendix B, I present the equation for the minimum lifetime of the heavier class of stars that end their lives as supernovae. (The Sun is not in this class, being less massive and longer-lived). Other than arbitrary constants that simply define the units one is using, this minimum lifetime depends on just three parameters: the strength of the electromagnetic force, α; the mass of the proton, m_p; and the mass of the electron, m_e. The relative strength of the gravitational force is reflected in the mass of the proton.

I find that long-lifetime stars that could make life more likely will occur over a wide range of these parameters. For example, if we take the electron and proton masses to be equal to their values in our universe, an electromagnetic force strength having any value greater that its value in our universe will give a stellar lifetime of more than 680 million years. The strong interaction strength does not enter into this calculation. If we had an electron mass 100,000 times lower, the proton mass could be as much as 1,000 times lower to achieve the same minimum stellar lifetime. This is hardly fine-tuning.

Of course, many more constants are needed to fill in the details of our universe. And our universe might have had different physical laws. We have little idea what those laws might be; all we know are the laws we have. Still, varying the constants that go into our familiar equations will give many universes that do not look a bit like ours. The gross properties of our universe are determined by these four constants, and we can vary them to see what a universe might grossly look like with different values of these constants.

I have examined the distribution of stellar lifetimes for one hundred simulated universes in which the values of the four parameters were generated randomly from a range five orders of magnitude above to five orders of magnitude below their values in our universe, that is, over a range of ten orders of magnitude.[14] While a few are low, most are high enough to allow time for stellar evolution and heavy element nucleosynthesis. Over half the universes have stars that live at least a billion years. Long stellar lifetime is not the only requirement for life, but it certainly is not an unusual property of universes.

I do not dispute that life *as we know it* would not exist if any one of several of the constants of physics were just slightly different. Additionally, I cannot prove that some other form of life is feasible with a different set of constants. But anyone who insists that our form of life is the only one conceivable is making a claim based on no evidence and no theory.

THE QUINTESSENCE OF DUST

While the topic is also somewhat technical, I think it is worthwhile to single out for additional discussion the one anthropic coincidence that is found to be the most puzzling and the most difficult to explain. It is one of the prime examples used by Ross and other theists when they promote the fine-tuning argument for the existence of God. This is what is known as the *cosmological-constant problem*.

When Einstein formulated general relativity around 1915, his equation for the curvature of space at a given point contained a constant term that was not constrained to any particular value. This was the *cosmological constant*. When positive in value, it provides for an effective gravitational repulsion. At first, Einstein included this term to balance the more familiar attractive gravitation to provide for the stable firmament of stars that was assumed at the time. When, a few years later, Hubble discovered that the universe was expanding, Einstein dropped this term from his equations, calling it his "greatest blunder." Actually, his greatest blunder was calling this a blunder.

The cosmological constant is often referred to as a "fudge factor," the implication being that it was something Einstein arbitrarily stuck into his equations to get an answer he wanted. This is somewhat misleading. The constant is required by those equations unless additional assumptions are made to rule it out. So, Occam's razor, which is a logical tool for determining the minimum requirements of an explanation, would require that the cosmological constant be included unless the data indicate otherwise. No fundamental basis has yet been found for taking the cosmological constant to be identically zero, although it is predicted to be zero by a principle called *supersymmetry* that many physicists think is a fundamental symmetry of nature. Until very recently, observational data indicated that the cosmological constant was very close to zero, and so this was what was assumed.

Cosmologists began to talk about a nonzero cosmological constant again in the 1980s, with the introduction of inflationary cosmology. An exponential expansion of the early universe was seen to follow from Einstein's equations in the absence of matter or radiation, with the curvature of space given by the cosmological constant.

More recently, unexpected evidence of the apparent current accelerating expansion of the universe has come from two independent studies of distant supernovae.[15] A nonzero cosmological constant has been considered as a possible explanation for this finding. Although this interpretation of the data is still preliminary at this writing, the universe today appears to be undergoing another round of inflation—much slower than the first.

Current observations indicate that the mass/energy of the universe is shared among its various components as shown in table 5.1.[16]

Table 5.1. The Energy Budget of the Universe

Radiation	0.005 %
Ordinary visible matter	0.5 %
Ordinary nonluminous matter	3.5 %
Exotic dark matter	26 %
Even-more-exotic dark energy	70 %

Neither the dark matter nor the dark energy has yet been identified. The existence of dark matter has been known about for some time, detected indirectly by its gravitational effect on the behavior of visible bodies such as stars. In just the past few years, however, the data have become good enough to determine that the amount of dark matter is insufficient to flatten the universe.

A geometrically flat or Euclidean universe is required by the inflationary model, which we can easily see as follows: The universe expanded by many orders of magnitude during inflation. After inflation, the space within our visible horizon is like a tiny patch on the surface of a balloon that has been blown up to a huge size. That patch will be very flat.

With observations in the mid-1990s indicating insufficient dark matter to flatten the universe, it was beginning to appear that the inflationary model might be wrong. This proved a real puzzle for cosmologists because the independent data coming in on the cosmic microwave background was providing increased support for the flat universe predicted by inflation.

Inflationary cosmologists have been rescued by the observed accelerating expansion and the inferred dark energy, which seems to be just sufficient to give a flat universe. Yet another gap for God to act in may have been closed. Read on, and we might be able to close one more.

What could this new dark energy possibly be? One possibility is that it is the result of a residual cosmological constant left over from the end of inflation. However, the prospect of a nonzero cosmological constant leads to an enormous difficulty that was recognized well before these latest observational developments. The cosmological-constant term in Einstein's equation is equivalent to a field with negative pressure and positive, constant energy density. As the universe expands, the total energy contained in the cosmological term will increase. In the time since the end of inflation, during the almost-linear big bang expansion, the cosmological energy would have increased by something like 120 orders of magnitude.

Currently, 13 billion years later, the dark energy is of the same order of magnitude as the other main components in table 5.1. This implies that it was "fine-tuned" at the end of inflation to be 120 orders of magnitude below what it is now. If, for example, the dark energy was just a hair larger at the end of inflation, that energy would be so great today that space would be highly curved, and the stars and planets could not exist.

This fact has not been lost on those theists, such as Ross, who see the hand of God in assuring that life, as we know it, could exist by fine-tuning the cosmological constant. However, recent theoretical work has offered a possible explanation for a nondivine solution to the cosmological-constant problem.

Several theoretical physicists have proposed models in which the dark energy is not the result of a cosmological constant at all but rather a dynamical energy field that does not have constant energy density. As a result, it evolves along with the other matter/energy fields of the universe and so need not be fine-tuned. The proposed field has been given the grand name of *quintessence*, after Aristotle's aether. In these models, the cosmological constant is exactly zero, as predicted by supersymmetry. Since zero multiplied by 10^{120} is still zero, we have no cosmological-constant problem in this case.

While the work on quintessence is highly preliminary and may not turn out to provide a viable explanation for the cosmological-constant problem, it is sufficiently interesting to mention at this juncture. If nothing else, it demonstrates that science is always at work trying to solve its puzzles. Furthermore, history shows that it has, so far, always succeeded in doing so within a materialistic framework. The assertion that God can be seen by virtue of his acts of cosmological fine-tuning, like intelligent design and all the earlier versions of the argument from design, is nothing more than yet another variation on the same old God-of-the-gaps argument. These rely on the faint hope that scientists will never be able to find an explanation for one or more of the puzzles that currently have them scratching their heads and will have to insert God into the remaining gaps.

NOTES

1. James Glanz, "Bang, You're Alive! On the verge of Re-Creating Creation. Then What?" *New York Times Week in Review*, January 28, 2001.

2. Richard Dawkins, *Unweaving the Rainbow: Science, Delusion, and the Appetite for Wonder* (Boston and New York: Houghton Mifflin, 1998), p. x.

3. The term "plasma" in physics refers to an ionized gas and is not to be confused with blood plasma.

4. For my narrative of the development and significance of these theories, see Victor J Stenger, *Timeless Reality: Symmetry, Simplicity, and Multiple Universes* (Amherst, N.Y.: Prometheus Books, 2000).

5. While physicist Alan Guth is usually given the primary credit for the idea of inflation, A. Guth, "Inflationary Universe: A Possible Solution to the Horizon and Flatness Problems," *Physical Review* D23 (1981): 347–56, several other physicists had come up with it at about the same time: D. Kazanas, "Dynamics of the Universe and Spontaneous Symmetry Breaking," *Astrophysical Journal* 241 (1980): L59–63; Andre Linde, "A New Inflationary Universe Scenario: A Possible Solution of the Horizon, Flatness, Homogeneity, Isotropy, and Primordial Monopole Problems," *Physics Letters* 108B (1982): 389–92. Nevertheless, Guth's book, *The Inflationary Universe* (New York: Addison-Wesley, 1997), is a good place to learn about inflation.

6. Recently, an alternative to inflation has been proposed called "the ekpyrotic universe," Justin Khoury, Burt A. Ovrut, Paul J. Steinhardt, and Neil Turok, "The Ekpyrotic Universe: Colliding Branes and the Origin of the Hot Big Bang," in *Physical Review* D64, no. 12 (2001): art. 123522. While current data cannot distinguish between the two models, the authors propose that future measurements of the polarization of the cosmic microwave background and gravitational waves should do so. We will have to wait and see. It should be noted that some of the existing observations that this new theory explains were built into its formulation, whereas in the case of the inflationary model, they were part of its predictions.

7. Charles L. Bennett, Gary F. Hinshaw, and Lyman Page, "A Cosmic Cartographer," *Scientific American* (January 2001): 44–45.

8. Stephen W. Hawking, *A Brief History of Time: From the Big Bang to Black Holes* (New York: Bantam, 1988), p. 129.

9. For general relativity purists, R here is the scale factor of the universe. Also, I should note that both Hawking and I have oversimplified

the question of the total energy of the universe from the purist's standpoint. However, the basic conclusion that zero energy was required to produce the universe is unchanged by a more sophisticated analysis.

10. The radioactivity of Earth also contributes to its temperature. In fact, life as we know it would not be possible without this added heat. I am neglecting this here for simplicity's sake, since this does not change the conclusion.

11. Since the CMB is not in thermal equilibrium, the photons it receives from stars and planets can go to increasing its temperature (although, as I have shown, this is negligible) and does not have to be reabsorbed by stars and planets. The CMB photons that are absorbed by Earth do not appreciably increase the entropy of Earth.

12. For example, the universe has a zero value of linear momentum, angular momentum, electric charge, and any of the other quantities that physics says are conserved, that is, neither created nor destroyed in physical processes.

13. Hugh Ross, *The Creator and the Cosmos: How the Greatest Scientific Discoveries of the Century Reveal God* (Colorado Springs: NavPress, 1995), p. 133.

14. Victor J. Stenger, "Natural Explanations for the Anthropic Coincidences," *Philo* 3 (2000): 50–67.

15. A. Reiss et al., "Observational Evidence from Supernovae for an Accelerating Universe and a Cosmological Constant," *Astronomical Journal* 116 (1998): 1009–38; S. Perlmutter et al., "Measurements of Omega and Lambda from Forty-two High-Redshift Supernovae," *Astrophysical Journal* 517 (1999): 565–86.

16. Jeremiah P. Ostriker and Paul J. Steinhardt, "The Quintessential Universe," *Scientific American*, January 2001, pp. 46–53.

SEVEN

THE OTHER SIDE
OF TIME

That natural reality is assumed rather than explained, is not proof for the existence of a creator. Introducing god as an explanatory notion only shifts the locus of the question: why would such a god exist? And, it is possible that the universe just happens to exist, without explanation.

—Willem B. Drees, 1955

THE BIG BANG AND CREATION

In the previous chapter, we found that no known laws of physics were necessarily violated at the supposed creation of the universe. The emergence of our universe from a preexisting, zero-energy emptiness would not have violated the principle of energy conservation, also known as the first law of thermodynamics. Neither would it have violated conservation of momentum, angular

momentum, electric charge, or any of the other conservation principles of physics that have been established by laboratory studies.

Likewise, the materialization of the universe from a prior state
of complete disorder, that is, maximum entropy, would not have
violated the second law of thermodynamics—a fact that may be of
great surprise to many. Even some scientists get confused on this
point, since it would seem that the second law then prevents order
from forming. In fact, as I showed, this is perfectly possible in an
expanding universe. While the total entropy of the universe
increases with time, as required by the second law, it could have
been as large as possible near $t = 0$ and still be less than maximum
now. This follows from the expansion of the universe, which
allows increasing room for order to form as time progresses and
the maximum allowable entropy increases faster than the actual
entropy.

As mentioned in chapter 3, the big bang has been regarded by
some theists as a vindication of their concept of a supernatural creation occurring sometime in a finite past. Recall the line of reasoning of Pope Pius XII: "Creation took place in time, therefore
there is a Creator, therefore God exists."[1] Hugh Ross has also
asserted that the big bang "proves" the truth of creation:

> The universe and everything in it is confined to a single, finite
> dimension of time. Time in that dimension proceeds only and
> always forward. The flow of time can never be reversed. Nor can
> it be stopped. Because it has a beginning and can move in only
> one direction, time is really just half a dimension. The proof of
> creation lies in the mathematical observation that any entity con-
> fined to such a half dimension of time must have a starting point
> of origination. That is, that entity must be created. This necessity
> for creation applies to the whole universe and ultimately every-
> thing in it.[2]

While Ross and other Christians claim the big bang provides
unique support for the tales found in their scriptures, it should be
noted that most of the thousands of religions practiced by humans
throughout history incorporate creation myths of one kind or

another. With a few exceptions, these superficially resemble the big bang—some even more so than the Bible! The Samhyka of India see space as condensing from the void. Similarly, an ancient Chinese view holds that the elements were produced in a mist of chaos. These sound a lot closer to modern cosmology than Genesis, which has Earth being created before the stars and contains a number of other statements that conflict with current knowledge.

Most Christians, Jews, and Muslims regard the creation stories of other religions as mythological. With regard to their own scriptures, opinions range from strict literalism to the view that the particular cosmogonies, as presented in the Bible or Koran, are the imaginary tales of a prescientific age that serve mystical rather than scientific purposes. Indeed, it takes quite a heavy massaging of the scriptures to make them conform to scientific wisdom. Let me discuss one example of such massaging in detail.

MASSAGING THE BIBLE TO FIT THE DATA

In *The Science of God: The Convergence of Scientific and Biblical Wisdom*, physicist Gerald L. Schroeder has attempted to show how the Bible can be interpreted as being consistent with modern cosmology.[3]

The Bible says God created the universe in six days and indicates the passage of only about 6,000 years since then, seemingly contradicting the estimate of 13 billion years from current cosmology. Schroeder argues that the six days of creation in the Bible refer to a different measure of time. He explains: " There is no possible way for those first six days to have an Earth-based perspective simply because for the first two of those six days there was no Earth."[4]

Instead, time during this six-day period was measured on a cosmic clock based on the vibrations of light (electromagnetic waves). Today the light from "creation" appears as the cosmic microwave background. The frequency of vibrations of this light has decreased by a factor of a trillion from the early universe period when atomic nuclei first began to form. Thus, according to Schroeder, the cosmic clock at that epoch ran off a trillion days for each of our modern days. The six cosmic days of creation then took about 16 billion years Earth-time. So, Schroeder claims, Genesis is not only consistent with cosmology, it gives the correct age of the universe, give or take a few billion years!

Each of the six days in Schroeder's Genesis actually takes a different length of Earth-time. Cosmic day one is 8 billion Earth-years long, and you divide by two to get the duration of each succeeding cosmic day. After six cosmic days occupying almost 16 billion Earth-years, we reach the time of Adam.

Cosmic day one starts 15.75 billion Earth-years ago and covers the creation of the universe, light "breaking free" as electrons bind to atomic nuclei, and the beginning of galaxy formation. This is described in Gen. 1:1–5 as the creation followed by light separating from the darkness.

Cosmic day two starts 7.75 billion Earth-years ago and lasts 4 billion Earth-years. During this period the stars and galaxies are born. This corresponds to Gen. 1:6–8, the formation of the heavenly firmament.

Cosmic day three starts 3.75 billion Earth-years ago. During 2 billion Earth-years, Earth cools, water appears, and the first life-forms appear. In Gen. 1:9–13, vegetation first appears during the third day.

Cosmic day four starts 1.75 billion Earth-years ago and lasts a billion Earth-years. Earth's atmosphere becomes transparent and photosynthesis produces an oxygen-rich atmosphere. Schroeder says that this corresponds to Gen. 1: 14–19 when "the Sun, Moon, and stars become visible in the heavens."[5]

Cosmic day five starts 750 million Earth-years ago and lasts 500 million Earth-years. During this period, the first multicellular animals appear and the oceans swarm with life. Gen. 1:20–23 says the waters bring forth swarms of living creatures and "birds fly above Earth."

Cosmic day six starts 250 million Earth-years ago and ends at the time of Adam. During this period we have a massive extinction in which 90 percent of life is destroyed and then repopulated with humanoids. This, Schroeder says, corresponds to what is described in Gen. 1:24–31.

Technically, Schroeder's formula gives the present as the end of the sixth day. However, it could just as well have ended a few thousand years ago and not affect the rest of the calculation where things are rounded off at hundreds of millions of years. Schroeder argues that after the six cosmic days of creation, Genesis switches its focus over to humanity and starts measuring time in human terms. The rest of the Bible concerns itself with the 6,000 Earth-years since Adam and Eve, estimated from the Bible in Bishop Ussher fashion.[6]

Schroeder does not deny the existence of hominid creatures before Adam. He talks about Neanderthals and Cro-Magnons, and accepts that they had developed tools, pottery, and many human-like qualities. In Lev. 11:33 the Bible mentions pottery. But, Schroeder argues that since it never refers to the *invention* of pottery, that event must have predated Adam.[7]

According to Schroeder, the Bible has no interest in these pre-Adam hominids because they were not yet fully human and had no souls. Thus they never appear. Adam represents the quantitative change to a large brain, but more important, the qualitative change that makes us different from all other forms of life: "our soul of human spirituality."[8] God breathed this into Adam, the first real human, 6,000 years ago.

Schroeder's attempt to connect thirty-one lines of Genesis to big bang cosmology and Earth paleontology makes entertaining reading. Let us return to the beginning and discuss it more critically. Schroeder's use of the formation of atomic nuclei as the defining moment for his cosmic timescale is completely arbitrary. He seems to have chosen it for no better reason than it gives the answer he wants, namely a factor of a trillion by which to slow his cosmic clock. Multiplying six days by a trillion gives 16 billion years, which is roughly consistent with our current estimate for the age of the universe, although about 3 billions years higher than the current best estimate.

Alternatively, Schroeder might have chosen some other moment in the early universe, say the time when radiation separated from matter. Indeed, he relates this event to the separation of the "light from the darkness" described in Genesis day one, so this choice would have made more sense. But, he already knows the answer he wants and sees this will not work. The fractional change in the vibration frequency of light from that moment to the present day is only on the order of one thousand, which would give an Earth-time interval of only fifteen years for the six cosmic days of creation. He needs it to be a million times longer. If he had chosen some other moment, he would have obtained yet a different timescale. In other words, Schroeder had many possibilities for a timescale and simply picked one that fit the best to modern cosmology.

And so, to give the right answer, Schroeder defines the beginning of the biblical universe to correspond to the time of quark confinement. That is not the beginning of the cosmological universe. Using his same method for relating time to the frequency of light, one can blueshift back from that point just as well as one can redshift ahead.[9] Thus, the events prior to quark confinement would recede infinitely into the past, in Earth-time, and we would have no creation at all!

Schroeder's use of an exponential function (doubling each day) to give different Earth-periods for each cosmic day is not justified by his argument that Earth-time is simply redshifted cosmic time. While an exponential relationship would apply for the inflationary epoch in the early universe, that was long ended by the time of quark confinement. After inflation, we have the Hubble expansion in which the redshift varies roughly linearly with time, not exponentially. By having each cosmic day half as long as the preceding one in Earth-years, again a completely arbitrary, unjustified procedure, Schroeder is able to vaguely relate events known from cosmology to those described in Genesis. However, even if we allow the arbitrary retrofitting of events to match the script, there is still much that is out of sequence.

In cosmic day two, 7.75 billion Earth-years ago, the "firmament" is created. Note that Schroeder thereby excludes from his firmament all galaxies more than 7.75 billion light-years away, which is most of the observable universe! Furthermore, he sees no problem with calling our expanding, evolving universe a "firmament." He selects his data carefully, accepting only those which agree with his hypotheses and discarding those which do not.

Primitive life first appears, in Schroeder's scheme, in cosmic day three. Here again it takes some mighty stretching to associate what is described in the Bible for the third day, including fruit trees, with the primitive life described by paleontology for that epoch.

Schroeder has the Sun, Moon, and stars becoming visible in cosmic day four. In fact, Genesis says that the Sun, Moon, and stars were created in this order of sequence—after Earth.

Cosmic day five has the waters teeming with life. But the biblical verses imply birds as well. Schroeder says that "birds" is a mistranslation and that the Bible here is referring to water insects instead. Translation is so easy when you know what you want a passage to say.

Cosmic day six contains the mass extinctions of life that occurred 65 million years ago. The biblical verses referenced make no mention of mass extinction. The biblical flood occurs well after Adam, but Schroeder needs to end the six days of creation with Adam for other purposes. This is one event he simply cannot make fit, although he is not honest enough to say so and leaves the impression that everything is consistent.

At first, the reader may get the impression that this book is a parody, with quite a few good chuckles when read in that context. However, the sections on evolution soon reveal that no parody is intended. They are just too unfunny, too formulaic. Schroeder trots out all the old, well-refuted design arguments about why "life could not have started by chance" and how the simplest forms, even viruses, are "far too complex to have originated without there being an inherent chemical property of molecular self-organization and/or reaction-enhancing catalysts at every step of their development."[10] He applies the usual creationist deception of calculating chance probabilities as if chance is the only operative mechanism, and then says this "proves" that God intervenes along the way when, unsurprisingly, they come out very low. Furthermore, Schroeder insists, the "staccato aspect of the fossil record" refutes classical evolution. "These rapid changes cannot be explained by purely random mutations at the molecular-genetic level.[11] The "staccato aspect" is called "punctuated equilibrium" by evolutionary biologists, and it is largely accepted as at least part of the pattern of evolutionary history and fully consistent with standard Darwinism.[12]

It takes quite a stretching of the imagination as well as a stretching of the timescale to make the biblical story of creation match that of modern cosmology, biology, and paleontology. In showing us how to make six days last 16 billion years, and evolution to agree with Genesis, Gerald Schroeder has demonstrated a plenitude of imagination.

The fact remains that, despite some crude similarity between the big bang and most creation myths, sacred texts describe a universe bearing little resemblance in detail to the one revealed by our telescopes. For example, the Bible describes an Earth-centered firmament of stars that is unchanged since creation. This sharply differs with astronomical observations, which indicate a centerless universe that has been evolving with time for 13 billion years. The light seen from galaxies far, far away left there long, long ago. Those galaxies look markedly different in detail from those observed closer to our time, say only a million years ago, although they exhibit the same fundamental physics.

Since the Copernican revolution, most biblical literalists have forgotten that the Bible is quite explicit in insisting that Earth is fixed in space. Henry Morris, for example, who insists that every word of the Bible is true, conveniently forgets this fact. However, he draws the line at the big bang—arguing that it did not happen.[13]

Theistic scientists like Ross and Schroeder accept the big bang and the rest of cosmology, seeing them as a confirmation of scripture. Biblical literalists like Morris sharply disagree. They reject the big bang and modern cosmology, because, in fact, it does conflict dramatically with scripture. Ross's and Schroeder's approach is called *apologetics*, while Morris makes no apologies. In the first case we have physicists who accept the results of science, and then attempt to bend and twist what is in the Bible to fit scientific data. In the second case we have an engineer who will have none of this bending and twisting. For Morris, the Bible is the immutable word of God, and if it conflicts with science, then science must be wrong. Of course, neither extreme is willing to seriously consider another possibility—that unbent and untwisted science shows simply and unequivocally that an unbent and untwisted Bible is grossly wrong in its picture of creation. It follows, by implication, that the Bible is highly suspect in its other doctrines as well.

BIG BANG SKEPTICISM

For many years, an alternative to the big bang, the steady-state universe of Fred Hoyle, Hermann Bondi, and Thomas Gold remained viable.[14] However, this particular model and others like it are now safely ruled out by observations, such as the evolution of galaxies mentioned above and the great quantitative and predictive success of the inflationary big bang model. Still, as we will see below, a version of the steady-state cosmology not envisioned by its originators may be resurrected in a model of multiple universes.

The big bang is occasionally called into question in the popular literature, such as in the 1991 book *The Big Bang Never Happened* by science writer Eric Lerner.[15] Lerner's critique is easily countered.[16] I know of no active contemporary cosmologist who takes his alternative plasma universe seriously. You can attend every conference on cosmology held these days and not find any sign that the big bang is in trouble.

Still, respectable but aging skeptics, such as astronomers Fred Hoyle (recently deceased), Geoffrey Burbidge, and Jayant Narlikar, and retired *Nature* editor John Maddox, can still be heard occasionally speaking out against the big bang.[17] Like the handful of late-nineteenth-century chemists and physicists who refused to their dying days to accept the atomic model of matter, long after it was solidly confirmed, big bang skepticism can be expected to rapidly diminish as the current holdouts pass from the scene.

The new creationists who use the big bang in support of their beliefs have exploited this fringe of dissent in cosmology, making it seem more representative than is the case and suggesting that objections to the big bang are theological rather than scientific. Indeed, a reading of Lerner would suggest this interpretation to someone not familiar with the field. Theistic physicist Ross asserts that general relativity and the big bang "prove a formidable threat to rational atheism."[18] Not for any rational atheists I know.

Let me assure you that mainstream cosmologists are not involved in any Vatican-led conspiracy to promote a creation cosmology. Most in my acquaintance are nonbelievers who see no sign of a creator in their data. They accept the big bang as well-established science, not theology. In any case, the truth is not determined by democratic vote, and most cosmologists are completely committed to letting the observations decide. They know full well that they would be disgraced, their careers ruined, if they were to allow religion or politics to influence their scientific judgments.

THE KALÄM COSMOLOGICAL ARGUMENT

Right now, observations strongly support the big bang. But whether they support the notion of a supernatural creator is another matter. One particular theological line of reasoning, of Islamic origins, can often be heard these days. This is known as the *kaläm cosmological argument*. Christian philosopher and professional debater William Lane Craig has presented it as a syllogism:[19]

1. Whatever begins has a cause.
2. The universe began to exist.
3. Therefore, the universe has a cause.

For the detailed philosophical arguments on both sides of the issue, see the book by Craig and philosopher Quentin Smith.[20] I will not repeat all the arguments here but focus on the main points and a counterargument of my own creation.

Craig gives no real justification for (1) calling it "intuitively obvious."[21] Perhaps it is intuitively obvious to him, but not to physicists like myself. Quantum phenomena, such as atomic transitions and radioactive decay of nuclei, seem to happen without prior cause. In fact, the highly successful theory of quantum mechanics does not predict the occurrence of these events, just their probabilities for taking place. While we leave open the possibility that causes may someday be found for such phenomena that allow for their prediction, we have no current basis for assuming such causes exist. After all, quantum mechanics is almost a century old and has been utilized with immense success over that period, with no sign of such causes ever being found.

Even if quantum processes are random, the creationist will still argue that they remain causal in nature. "Where did the laws of chance come from?" they might ask, imagining God playing dice. And, "Where did quantum mechanics come from?"

Of course, the atheist should ask the theist to answer the same questions. When he or she gives the expected response, "They came from God," the atheist can then retort, "Where did God come from?" However, Craig and other theologians are well aware of this response and argue that, being eternal, God did not have a beginning and so did not require a cause.

Like many of the arguments discussed in this book, it all depends on who is forced to carry the burden of proof. Polished debaters like Craig work very hard on stage to pass the burden off on their opponents—who are frequently less skilled at the game and easily fall for the ploy. However, in this case, Craig and his theist colleagues cannot cast off their burden with clever rhetoric alone. Their theism is the less parsimonious hypothesis, requiring something more than the purely natural to explain phenomena. The fact that we have, with conventional quantum mechanics, an example of a noncausal mechanism is sufficient to refute the kalām premise (1) that everything that begins has a cause. And since the argument is presented as a syllogism, this is sufficient to refute the conclusion that the universe must have had a cause.

The kalām cosmological argument is yet another form of the argument from design, and we have seen how many times this has risen from the dead. So let me proceed to hammer some more nails into kalām's coffin—perhaps even drive a stake into its heart.

For Craig, the empirical evidence for the big bang justifies premise (2), which asserts that the universe had a beginning. Craig also makes an elaborate philosophical and mathematical argument, in essence concluding that an infinite regress into the past cannot occur, and so time must necessarily have a beginning. Here, he seems to assume some kind of Platonic reality to time. However, if we use the realist physicist's operational definition of time as the number of ticks on a clock, then we can have a denumerable infinity of time in the past just as well as the future. That is, we can think of time as a counting process that can continue indefinitely into the future or the past. In fact, as I explain below, the principles of physics indicate a symmetry in time at the fundamental level and so what holds for the future must also hold for the past.

Previous responses to Craig, by Smith and others, have not disagreed with premise (2) per se, but questioned whether it even made any sense to talk about a cause before the existence of time. A common assumption among theist and atheist philosophers alike, following St. Augustine, is that time started at the beginning of the universe. This is usually the position taken by the atheists who debate Craig and other theists. Indeed, I also presented this view in my 1988 book, *Not By Design*. While I do not question that this is still a viable position, another position exists that cannot be ruled out within the framework of current knowledge.

I have proposed an alternative response to kalām premise (2) in which the assumption of a beginning to time (though not the big bang itself) is disputed.[22] I have shown that the universe did not necessarily have a beginning and that the time which we identify as the beginning, call it $t = 0$, is an arbitrary point. Time and space exist on both the negative and positive sides of the time axis.[23]

Recall that the kaläm argument holds that since the universe had a beginning, it must have had a cause. God, on the other hand, had no beginning and so required no cause. It follows that if I can demonstrate the universe need not have had a beginning, then Ross, Craig, and other theists who use the kaläm argument will be hoisted on their own petard. If the universe did not necessarily have a beginning, then it did not necessarily have a cause.

THE ARROW OF TIME

Most people would agree with Ross when he states that "the flow of time can never be reversed." Certainly this relates the personal experience of all of us. Anyone who suggests otherwise is likely to be assumed to be either a science fiction writer or somewhat daft. I am not a science fiction writer, and I hope the reader will reserve judgement on my sanity until I have had a chance to explain why I think that this commonsense belief does not apply at the fundamental level of reality.[24]

First, none of the basic principles of physics includes a preference for one direction in time over the opposite direction. This was true for classical Newtonian physics, and it remains true for relativistic and quantum physics. Some very rare elementary particle processes exhibit a small, one part in a thousand, preference for one time direction over the other. However, they do not forbid time reversal, just perfect time symmetry. All other fundamental processes occur with the same probability in either time direction. For example, in chemistry you can have the reaction $H_2 + O \rightarrow H_2O$ + energy or, its reverse, energy + $H_2O \rightarrow H_2 + O$. In physics, a pendulum can swing up as well as down. When you watch a film of a fundamental physical or chemical process (with the one exception mentioned above), you cannot tell from just looking at the screen whether the film is being run forward or backward through the projector.

Then, you may ask: If this is the case, why is it that we experience a profound, singular direction of time? If time can run either way, why does it run just one way for us? Why do we never grow younger? Why doesn't a broken glass reassemble? Why doesn't a dead man rise?

The answer is: All these thing are possible! Just highly unlikely in the age of the universe. As described in chapter 4, over a century ago physicist Ludwig Boltzmann proved that when you have an isolated system containing many particles moving about more-or-less randomly, then that system will tend to an equilibrium state in which the entropy, that is, disorder, of the system maximizes. This has the appearance of the second law of thermodynamics, which, we have seen, says that an isolated system becomes more disorderly as time moves forward. Our personal experience, with our own bodies becoming more disorderly as we age, confirms this notion.

Boltzmann noted, however, that the direction of time—what Sir Arthur Eddington later dubbed the *arrow of time*—can be regarded as being defined as the direction of increasing entropy. That is, rather than assuming, as we usually do, that time has a fundamental direction, we can drop this assumption. Then both directions are possible, and we simply and arbitrarily label as the "future" the direction of increasing entropy. In other words, the arrow of time is just a statistical definition. Although many scientists since Boltzmann have sought to explain the arrow of time in terms of nonstatistical principles, none have succeeded in doing so.

Although it seems to violate common experience, time reversibility is not that hard to understand. Consider what happens when you punch a hole in an inflated tire. Air molecules inside rush out until the pressure inside is the same as the pressure outside and equilibrium is reached. No matter how long you wait, the time-reversed process in which molecules of air rush through the puncture to reinflate the tire never seems to happen. So we think of the arrow of time as pointing in the direction in which a tire deflates.

However, when we look at individual particles, it will occasionally happen that a molecule of air from the outside is moving in the right direction and enters the tire through the small opening. Or this might occur with several molecules. If the inflated tire contained only two or three molecules to begin with, then we could not say which direction of time is "forward" or "backward." The chances of them going into the tire would not be that much lower than the chances of them coming out.

Indeed, it is possible, in principle, that a trillion trillion molecules will all be moving at once in the direction of the hole, spontaneously inflating the tire. This is, of course, highly unlikely. But my point is that it is not strictly forbidden by any known principles of physics.

The arrow of time is not a fundamental property of the universe. It applies only to systems containing large numbers of particles where some processes, like aging, are very much more likely to occur in one direction than the other. Obviously, it serves as a useful concept that reflects human experience.

Second, when we look at the quantum world, the case for time reversibility becomes even more compelling. The main reason many quantum processes seem so weird is that they, in fact, appear to exhibit backward causality. In the quantum domain, experiments have demonstrated that the future has an effect on the past. No doubt this is counterintuitive, but so once was the fact that Earth is round. The human predisposition to think only in terms of time "flowing" in one direction has led some physicists and philosophers to cook up arbitrary and unnecessary schemes, like "hidden variables" and "many worlds," to account for the strange behavior of quantum events.

For example, quantum particles such as electrons or photons seem to act as if they are at several places at the same time. An electron in an atom moves instantaneously from one orbit to another. This seems to violate the principle, elucidated by Einstein in his 1905 special theory of relativity, that no physical body, or signal, can move faster than the speed of light. For a particle to be seen in two places at once, you would think that it must have moved with infinite speed between the two points.

However, it is possible for a fundamental particle to be two places at once without exceeding the speed of light. All it has to do after it passes one point is to go back in time and then forward again, passing a second point at the same time it passed the first one. This notion was implemented by Richard Feynman in 1948 when he explained the nature of antimatter. He showed that anti-electrons, or positrons, were simply electrons going back in time. Although most physicists still describe events in terms of the conventional arrow of time, time reversibility is today deeply embedded in fundamental physics theory.

WAS THERE A BEGINNING?

Now let me return to the question of whether or not the universe had a beginning. My scenario is provided by the inflationary model that, as we have seen, currently supplements big bang cosmology. The new creationists and I agree that the big bang is strongly supported by astronomical observations. As we saw in chapter 6, while inflation may still be tentative, it is the only current theory that successfully explains a wide range of observations. We also noted that it has already successfully passed several observational tests which might have falsified the model. I have already described the basic ideas, but let me briefly go over them again.

Inflation can be shown to follow from the equations of Einstein's general theory of relativity. This theory first appeared in 1915 and has since passed every empirical challenge, including several that could have demolished it. When Einstein's equations are written down for a space that is empty of matter or energy but has a nonzero cosmological constant, they allow for an exponential expansion on *both sides* of $t = 0$.

Normally the side of the time axis with negative t is ignored, since we are on the positive t side and we cannot look back through the chaos near $t = 0$ to see what went on before. However, although we cannot prove that anything was there, we cannot prove that nothing was. Note again the importance of the assignment of the burden of proof here. I am not saying that I can demonstrate with high probability that the universe existed before the big bang. My claim is only that well-established principles of physics do not rule it out and, indeed, the equations include such solutions. Thus, those who assert that nothing existed before the big bang have the burden of pointing out what well-established scientific principle demands that to be the case. If none, then they are making an additional, uneconomical hypothesis that is not required by the data and should be sliced cleanly away by an application of Occam's razor.

Now, from our point of view, the other side of $t = 0$ is in our deep past, exponentially *deflating* to the void prior to the quantum fluctuation that then grew to our current universe. However, from the point of view of any observers who might live in the universe at that time, their future is into our past—the direction of increasing entropy on that side of the time axis. These observers would experience a universe expanding into their future, just as we experience one expanding into our future. In other words, each side of the time axis has an arrow pointing away from the origin.

Would these different parts of the universe be identical—kind of mirror images of each other? Not unless physics is completely deterministic, which we do not believe to be the case. The two parts would more likely be two very different worlds, each expanding in its own merry way, filled with all the other random events that lead to the evolution of galaxies, stars, and perhaps some totally different kind of life.

This scenario also serves to explain why we experience such a large asymmetry in time on the cosmic scale, while our basic equations exhibit perfect time symmetry.[25] Fundamentally, the universe as a whole is time-symmetric, running all the way from minus eternity to plus eternity with no preferred direction, no arrow of time. Indeed, the whole notion of *beginning* is meaningless in a time-symmetric universe. And, without a beginning, the kaläm cosmological argument for a creator fails again, this time because of the failure of step (2) in Craig's syllogism.

The situation I propose here is not to be confused with the *oscillating universe* scenario that many authors have suggested in the past. That proposal was based on the possibility that the universe is gravitationally closed, that is, has negative total energy so that it eventually collapses on itself in what has been termed "the big crunch." Then, it is imagined, a new big bang starts all over again. In the oscillating universe, time still runs in one direction. I am suggesting that time's arrow, as defined by the direction of increasing entropy, runs one way on one side of $t = 0$ and the other way on the other side. In my scenario, we have an infinite, eternal, and symmetric universe with no beginning and, time-symmetrically, no end. The time we call $t = 0$ is simply the time at which the universe was as small as it could possibly be.

AT THE PLANCK TIME

Let us now explore what may have happened at $t = 0$. I want to emphasize again that I am not wildly speculating here but considering the implications of existing, well-established theoretical and observational knowledge. For this purpose, since human language is so heavily dependent upon tense, I will lapse back into the vernacular in which an arrow of time is taken as pointing in the conventional direction. This presents us no problem, as long as we remember that we can just as well use language in which the arrow points the other way.

Of course, we do not know for sure what happened at $t = 0$. All I can do, and all I need to do to make my point, is provide a scenario that does not violate any known principles and does not invent any new ones in order to make sure everything comes out in some preconceived fashion. My scenario need not be unique. Other scenarios may work equally well, but they should not be preferred over what I present unless they are at least equally parsimonious and have other arguments in their favor.

Let me begin by repeating that $t = 0$ is not a unique time, just a label for that arbitrary point on the time axis where the inflationary big bang begins. You will often read that we cannot understand what happens very near $t = 0$ because that requires a theory of quantum gravity, a marriage between quantum mechanics and general relativity which does not yet exist. This is true, but it is also true for any other point in time, like the instant the reader's eye reaches the period at the end of this sentence.

Every point on the time axis, including "now," is surrounded by a small time interval within which our current theory of gravitation, Einstein's general theory of relativity, does not apply. General relativity is not a quantum theory, and quantum gravitational effects are important when time intervals are as small as 10^{-43} second (43 orders of magnitude less than a second). As we saw in chapter 6, this is called the Planck time. Similarly, every point in space is surrounded by a tiny sphere of radius 10^{-35} meter, a distance called the Planck length, where again general relativity does not apply.

Now, the Heisenberg uncertainty principle of quantum mechanics makes it impossible to measure any time interval smaller than the Planck time or distance smaller than the Planck length. According to physics convention, time and distances are operationally defined by their measurements (see appendix A for technical details). Thus, unless one changes that convention, it is meaningless to talk about distance and time as having any qualitative values within a Planck sphere—even with a theory of quantum gravity. Furthermore, the inability to define distance and time within these dimensions implies that no other physical quantity can be measured, since they are all defined in terms of space and time. And if no measurements are possible, neither is information. Thus we have a condition of complete uncertainty, or maximum entropy.

And this was the state of the universe in the 10^{-43} second time-interval around $t = 0$, if it was confined within a Planck sphere as inflationary cosmology implies. The universe was in a condition of maximum entropy—total chaos. If a supernatural creation occurred at this point, it was a creation without design since the universe was without order.

So what did happen at t = 0, even if it was "not by design"? The common view among cosmologists today, although they may differ among themselves on the details, is that an uncaused *quantum fluctuation* led to the inflation which preceded the appearance of matter and structure. Edward Tryon may have been the first to publish this idea in a major scientific journal.[26]

With the appearance of the inflationary cosmology in 1981, several authors developed models in which inflation is triggered by an initial quantum event.[27] These papers may be too technical for the reader.[28] However, they should serve to illustrate that serious attention to the possibility of an uncaused origin of the universe has been given by reputable physicists and published in major physics journals. Although none claim to "prove" that the universe began in this way, if the notion was nonsense and violated any known principles of physics, these papers would never have been published. I will not discuss all the various models but simply try to illustrate the principles involved.

The uncertainty principle allows energy (which is equivalent to mass by $E = mc^2$) to appear spontaneously, as long as it disappears in a short-enough time. In the inflationary-universe scenario, the energy appears not as matter or radiation but in the curvature of space as expressed by the cosmological constant of general relativity. As we have seen, the solution of Einstein's equations in this case (the de Sitter solution) is then exponential. However, as we have also seen, the total energy of the universe is zero, so on a longer timescale energy conservation is not violated.

Now, if all this happened at $t = 0$, you may ask, why doesn't it happen at other times? Why hasn't another big bang occurred within our own universe in 13 billion years? Occasionally you will read a scare article in which the author warns that particle accelerators, like the one at Brookhaven National Laboratory, will produce a high energy density that will mimic those early universe conditions and may cause the universe to blow up! Not to worry. First, independent of any theory, cosmic rays in space have energies far exceeding those we have been able to generate artificially on Earth. These have not destroyed the universe yet, so we can rest assured that neither they nor scientists busy at work in their accelerators will do so in the near future.

Second, we can understand from theory why the universe does

not spontaneously explode. The exponential expansion that we associate with inflation occurs only under the condition in which the universe is empty of matter or radiation. Solutions of Einstein's equations are much more benign in the presence of matter and radiation, which constitutes the current situation in our universe. You can think of the gravitational pressure from that matter as holding off the explosion.

THE MULTIVERSE

Obviously, if our universe appeared as a quantum fluctuation in a space-time void, this could have happened more than once—and probably did. The multiple-universe scenario is implied by the original suggestion of Tryon and required in the chaotic inflationary model of Andre Linde.[29] If, as was argued above, the universe as a whole is infinite in extent in both space and time, then baby universes can be expected to pop up randomly at different positions and times. They appear as expanding bubbles that move away from one another, never colliding or coalescing.

While the multiple-universe scenario is not required to deflate the kaläm argument, which fails on several fronts, it can be used to provide a natural explanation for the so-called anthropic coincidences. As we have seen, the relatively new but now commonly asserted argument for a creator based on the fine-tuning of the constants of nature is also fallacious. While life *as we know it* would not be possible if one of several of these constants were slightly different, no one can demonstrate that *other forms of life* could not have occurred in a universe with different constants or laws. Now, I cannot prove that multiple universes exist or that other forms of life are possible. But, again, I insist that I do not have the burden of proof. That burden rests on the shoulders of those who make the fine-tuning argument. They must prove why multiple universes are impossible.

Just as kaläm can be challenged on several fronts, so can the fine-tuning argument. The multiple-universe scenario provides a simple mechanism by which universes of all types of structures can arise. Certain principles of physics, such as energy conservation and the laws of relativity, can be expected to apply to the "multiverse," since these can be shown to arise from the symmetries of the void (see chapter 8). This is what enables us to still talk about the multiverse in physics terms.

However, each baby universe will have, in addition to the global symmetries of the multiverse, certain *spontaneous broken symmetries* that will lead to its unique structure. These broken symmetries are accidental, so the structure of each universe can be expected to differ from the others. This means that the kinds of particles present, the forces between them, and various physical constants can be expected to be different from baby universe to baby universe. Some of the baby universes will likely contain little of interest, and some may not contain structures like stars that live very long and manufacture complex elements that can serve as platforms for life. But many others can be expected to contain complex systems capable of evolving into something resembling life (or, indeed, perhaps something resembling nothing with which we are familiar and even far exceeding human life and mind in wondrous capabilities). Thus, our universe only appears to be fine-tuned for us because it is that baby universe which happens to contain the properties needed for our kind of life to evolve.

Some theists have argued that the multiverse scenario is less parsimonious than one in which only a single universe exists. However, since multiple universes are suggested by existing knowledge, and no known principle rules them out, it becomes, in fact, less economical to assume a single universe. Occam's razor applies to the number of hypotheses of a theory, not the number of elements. Yet again, the person making the less parsimonious proposal has the burden of proof.

NOTES

1. Pius XII, "The Proofs for the Existence of God in the Light of Modern Natural Science," address of the pope to the Pontifical Academy

of Sciences, November 22, 1951, reprinted as "Modern Science and the Existence of God," *Catholic Mind* 49 (1972): 182–92.

2. Hugh Ross, *The Creator and the Cosmos: How the Greatest Scientific Discoveries of the Century Reveal God* (Colorado Springs: Navpress, 1995), p. 80.

3. Gerald L. Schroeder, *The Science of God: The Convergence of Scientific and Biblical Wisdom* (New York and London: Free Press, 1997).

4. Ibid., p. 51.

5. Ibid., p. 67.

6. James Ussher (d. 1656) was an Irish archbishop whose calculation, based in biblical chronology, that the creation took place in 4004 B.C.E. was long accepted.

7. Schroeder, *The Science of God*, p. 130.

8. Ibid., p. 133.

9. Recall that redshifting means decreasing the frequency or increasing the wavelength of light, as in the Doppler effect described in the text. Blueshifting is the opposite.

10. Schroeder, *The Science of God*, p. 85.

11. Ibid., p. 87.

12. Roger Lewin, *Complexity: Life at the Edge of Chaos* (New York: Macmillan, 1992).

13. Henry M. Morris, "Cosmology on Trial," BTG no. 137, Institute for Creation Research, El Cajon, California.

14. Fred Hoyle, "A New Model for the Expanding Universe," *Monthly Notices of the Royal Astronomical Society* 108 (1948): 372–82. H. Bondi and T. Gold, "The Steady-State Theory of the Expanding Universe," *Monthly Notices of the Royal Astronomical Society* 108 (1948): 252–70.

15. Eric J. Lerner, *The Big Bang Never Happened* (New York: Times Books, 1991).

16. Martin Gardner, "The Big Bang Theory Still Lives," *Skeptical Inquirer* 16, no. 4 (1992): 357–61; Victor J. Stenger, "Is the Big Bang a Bust?" *Skeptical Inquirer* 16, no. 4 (1992): 412–15.

17. Fred Hoyle, Geoffrey Burbidge, and Jayant V. Narlikar, *A Different Approach to Cosmology: From a Static Universe through the Big Bang towards Reality* (Cambridge: Cambridge University Press, 2000); John Maddox, *What Remains to Be Discovered?* (New York: Free Press, 1999).

18. Ross, *The Creator and the Cosmos*, p. 81.

19. William Lane Craig, *The Kaläm Cosmological Argument*, Library of Philosophy and Religion (London: Macmillan, 1979).

20. William Lane Craig and Quentin Smith, *Theism, Atheism, and Big Bang Cosmology* (Oxford: Clarendon Press, 1993).

21. Craig, *The* Kaläm *Cosmological Argument*, p. 141.

22. Stenger, *Timeless Reality: Symmetry, Simplicity, and Multiple Universes* (Amherst, N.Y.: Prometheus Books, 2000).

23. Philosophical note: By time and space "existing," here, I do not necessarily mean that they are in some sense substantial elements of reality. I simply mean that the concepts of space and time may still be meaningfully applied before the big bang.

24. A more complete presentation of these arguments can be found in Stenger, *Timeless Reality.*

25. For a good discussion of the philosophical issues concerning time symmetry, see Price, *Time's Arrow and Archimedes' Point: New Directions for the Physics of Time* (Oxford: Oxford University Press, 1996).

26. E. P. Tryon, "Is the Universe a Quantum Fluctuation?" *Nature* 246 (1973): 396–97.

27. David Atkatz and Heinz Pagels, "Origin of the Universe as Quantum Tunneling Event," *Physical Review* D25 (1982): 2065–73; S. W. Hawking and I. G. Moss, "Supercooled Phase Transitions in the Very Early Universe," *Physics Letters* B110 (1982): 35–38; Alexander Vilenkin, "Creation of Universes from Nothing," *Physics Letters* 117B (1982): 25–28; Andre Linde, "Quantum Creation of the Inflationary Universe," *Lettere Al Nuovo Cimento* 39 (1984): 401–405.

28. For a popular-level article on the subject, see Jonathan J. Halliwell, "Quantum Cosmology and the Creation of the Universe," *Scientific American* (December 1991): 76–85.

29. Andre Linde, "Chaotic Inflation," *Physics Letters* 129B (1984): 177–81; "Quantum Creation of the Inflationary Universe"; *Particle Physics and Inflationary Cosmology* (New York: Academic Press, 1990); "The Self-Reproducing Inflationary Universe," *Scientific American* 271, no. 5 (1994): 48–55.

EIGHT

THE LAWS OF THE VOID

Since you must admit that there is nothing outside the universe, it can have no limit and is accordingly without end or measure. It makes no odds in which part of it you may take your stand; whatever spot anyone may occupy, the universe stretches away from him just the same in all directions without limit.

—Lucretius, ca. 55 B.C.E.

THE PLATONIC ORDER

I have been developing the scenario by which our universe could have evolved naturally from an initial state of emptiness into the vast cosmos visible to us today, with its local pockets of order like Earth. Although certainly not proven, this picture is suggested by modern physics and cosmology. While this scenario may turn out not to be correct or need modification as our knowledge evolves, it at least represents a counterexample serving to refute the claims one hears frequently these days that current scientific knowledge necessarily points to a supernatural creation of the universe.

We have seen that zero external energy was required to produce the mass and energy of our universe. We have seen that order can spontaneously arise from disorder. We have seen that complexity can evolve from simplicity. We have seen that time has no fundamental arrow, and so the very concepts of cosmic beginnings and causal creation are problematical. No known scientific principles are necessarily violated in a model of our universe that is causally self-contained, in which everything that happens, happens within.

In this context, I need to make it clear that when I talk about "our universe," I mean all that we see with our unaided eyes plus what we see when we look into our most powerful telescopes and microscopes. It also includes what we infer from the less direct data gathered by our most sophisticated instruments in experiments of every variety, in all scientific fields. The possibility exists of other universes besides our own. Multiple universes are suggested by modern cosmological theories, although, since they cannot be observed, their scientific status may be questioned. However, nothing in our current knowledge rules them out, so we cannot simply eliminate them from all scientifically consistent speculation, especially when otherwise successful theories betoken their presence.

The notion of multiple universes, in this context, is not to be confused with the parallel universes that are often referred to in discussions of quantum mechanics. In the *many-worlds* interpretation of quantum mechanics,[1] all the different paths that a physical system may take are viewed as occurring, *in reality*, in different "worlds." Unlike the multiple universes suggested by modern cosmology, quantum many worlds are viewed as interacting with one another to produce observed quantum effects that appear puzzling or even bizarre. While the many-worlds interpretation has many adherents, and the associated mathematical formalism has great merit, other viable interpretations of quantum mechanics remain with no consensus on which, if any, is the correct one. In any case, this need not concern us here.[2]

Even in the context of other possible universes, I claim that ours shows evidence of being self-contained. If it appeared as a quantum fluctuation in a larger universe, that process was a non-causal one. Even if this were not the case and our universe was brought into being by some causal process such as the act of a creator god, the data indicate that it was once in a state of complete chaos and so currently retains no memory of that creative act. This conclusion is obtained by extrapolating the big bang back to the earliest definable moment within our universe. This moment, as we saw in chapter 7, is not $t = 0$ but $t \approx 10^{-43}$ second, the Planck time. At that time, our universe was confined within a sphere of radius 10^{-35} meter, the Planck length. According to the Heisenberg uncertainty principle of quantum mechanics, no smaller distances or times can be operationally defined (see appendix A), so no information exists inside a Planck sphere. Thus, our universe at this earliest time had maximum entropy (see appendix C). It was a kind of black hole that cannot be seen into and so exists in a state of maximum disorder. Even if the universe was created—and we have seen that nothing requires that it was—nothing in the "mind" of the creator could have carried through to even the first tiny fraction of a second after creation. If a creator plays any role in our universe, it must be through actions taken after the Planck time.

All of this is based on currently known principles of physics. Still, you are likely to ask: Where did the scientific principles themselves come from? In this chapter I will tackle this question and in the process further elaborate the model of a wholly natural and self-contained universe.

Theists, and even some nontheistic scientists, argue that the very existence of scientific principles themselves provides evidence for a Platonic order to the universe that transcends the realm of our observations. As physicist and Templeton Prize–winner Paul Davies has put it:

The very fact that the universe is creative, and that the laws have permitted complex structures to emerge and develop to the point

of consciousness—in other words, that the universe has organized its own self-awareness—is for me powerful evidence that
there is "something going on" behind it all. The impression of
design is overwhelming.[3]

Contemporary trends in liberal Christian theology and its supposed rapprochement with science have moved it closer to a pantheist position in which deity is to be found in the order of nature.
The Christian God is seen as a creative entity—transcending space,
time, and matter—that is ultimately responsible for all that exists.
And here is where some scientists and theologians currently seem
to share a common ground—in the notion that ultimate reality is
not to be found in the quarks, atoms, rocks, trees, planets, and stars
of experience and observation. Rather, that reality exists in the
mathematical perfection of the equations of physics and in the theological perfection of an entity existing, along with those equations, in a realm beyond human observation. This god is know-
able, not by his appearance before us, but by his presence as that
perfect reality. We all exist in the "mind of God."

Of course, the Christian concept of God has always been a
curious mixture of Hebrew, Greek, and Eastern mystical traditions,
among others. Perhaps that is the secret to its longevity. The gospel
writers shaped their stories of Jesus around the model of the god-
man of Mithraism and other Middle Eastern cults, with a little
Osiris and Dionysus thrown in. This helped make the new religion
attractive to the masses. Early church theologians, especially
Augustine of Hippo (d. 430), provided a philosophical model of a
transcendent God that was strongly reminiscent of Plato's Form of
the Good, though molded to fit the YHWH of Judaism. This
helped make Christianity intellectually respectable. So this combination—lots of magic, miracles, festivals, and rituals for the
common folk, combined with a rational theology for the intelli-
gentsia—has survived to the twenty-first century.

Until recently, the application of Platonism to science has not
been widespread, or at least not discussed much in the open. It
represents just the sort of talk, rather disconnected from the
observable world, that most scientists disdain. Today, however, we

find a number of theoretical physicists joining in the Platonic chorus. Although avowed atheists or agnostics, Stephen Hawking, Roger Penrose, Steven Weinberg, and other prominent theoretical physicists and mathematicians warble Platonic melodies of a reality manifested in the equations of mathematics and physics. They view *quantum fields* and space-time *metric tensors* as "more real" than quarks and electrons.

Many physicists have speculated that theoretical developments will eventually lead to a "theory of everything" (TOE) in which all the basic principles of physics will be shown to follow from a minimal set of assumptions, perhaps just statements of self-consistency. In that case, as Einstein suggested, there may have been nothing for God to do.[4] One route that is now being intensely explored is *string theory*, or its more recent extension called *M-theory*.[5] Since this theory is still far from being experimentally testable, we will have to wait and see. What I will try to show here is that, independent of M-theory or any other TOE, many of the most important principles of physics do indeed follow from little more than arguments of consistency, while others may be simply a product of chance.

In contrast to Platonic theorists, I am a mundane experimental physicist who sees reality as composed of objects that kick back at you when you kick them. Kick a rock with your foot, and it kicks back. Kick an atom or an electron with a photon, and it kicks a photon back to you. By measuring the properties of these photons, you can learn about the atom or electron. The properties of matter inferred from these measurements seem more real to me than metric tensors and quantum fields.

A reality of uncuttable atoms, that is, elementary objects (not to be confused with the "atoms" of chemistry) moving around in an otherwise empty void, as suggested by Leucippus and Democritus twenty-five centuries ago, is buttressed by the highly successful standard model of elementary particles and forces developed in the 1970s. This model continues to offer the simplest picture consistent with all the data. Tied in with the rest of physics, the standard model suggests a plausible, purely natural, explanation for the laws of physics. As we will see, universal laws, such as energy and momentum conservation, are simply mathematical statements about the natural symmetries of space and time. Other laws may have resulted from the spontaneous breakdown of those symmetries. Indeed, the laws of physics indicate that the universe looks as it does because it has the same basic properties as an empty void, just what would be expected if it appeared spontaneously, without cause or creation, out of the void.

THE PROPERTIES OF THE VOID

What are the properties of the void? The void, as I visualize it, is not "nothing" in the absolute, philosophical sense. If we take nothing to literally mean "no thing," then it cannot even be described. For if it is to be described, it must have some properties and, then, it is not *no thing* but *some thing*. Being an experimental physicist whose head spins at such thoughts, I will define "void" as that minimal arena that permits any experiment at all.[6]

Aristotle argued that the void cannot logically exist. However, any logical argument depends on its premises, and scientific questions are never settled by logic alone. Ultimately, science relies on observations. So let us approach the problem of the void as experimental physicists would. Imagine we are engaged in a research project entitled *An Experimental Investigation into the Properties of the Void*. This will be what Einstein called a *gedankenedperiment*, a thought experiment carried out in the mind and on paper.

In our thought experiment, we imagine a region of space in which we have removed all the matter and energy. What is left is the void. Now, it may be objected that this void is still "something" rather than "no-thing," so I must make it clear that I am hypothesizing the ontology mentioned above, and justified in *Timeless Reality*, in which reality is composed of "atoms" in the void and no more. When we remove all the atoms, then we are left with "nothing," since there is no other element of reality. Of course, this ontology may be wrong, but if I can show that it is sufficient to explain the data, then no one can claim that the universe, by its very existence, requires divine creation.

In our thought experiment, we attempt to measure the properties of the void as we attempt to measure the properties of anything else—by utilizing some measuring instruments. The most familiar instruments of the elementary physics laboratory are the meter stick for measuring distance and the clock for measuring time. Let us choose to use these here. Note that, in doing so, we are simply deciding to describe the void in terms of space and time. This does not necessarily imply that the object of our inquiry is in any sense composed of "substances" called space and time. Rather, we simply make use of the readings off a meter stick and a clock to describe the observed goings-on inside the void in a certain language, namely, the language of space and time.

The choice of space-time language is not the only possibility. We might choose to use a different language determined by measuring instruments other than meter sticks and clocks. In particle physics experiments done in accelerators, for example, we use instruments that measure the energy and momenta of whatever entities trigger their electronic circuits. This usually provides enough information to describe particle collisions without ever introducing space-time variables.[7] In fact, the theoretical equations that describe these interactions use energy and momentum as their independent variables, along with other quantities such as angular momentum, spin, electric charge, and many other variables. At one time, back in the 1960s, one school of particle physics sought to do away with space and time altogether. That hasn't happened, but maybe someday it will. Modern M-theory, mentioned above, is motivated by some of those ideas.

Nevertheless, we are all familiar with space and time, and these variables remain the most intuitive. So I will use space-time language in describing the void.

As our research proceeds, we might introduce other instruments besides meter sticks and clocks, but these still would be calibrated in terms of measurements of distance and time. In fact, even our meter stick is calibrated using a clock, the meter being defined by international convention as the distance light goes, in a vacuum, in $1/3 \times 10^{-8}$ second.

Of course, no objects exist inside our void for which we can measure the time or distance. However, we can follow a procedure that is common in physics when studying the properties of various "fields" such as gravity or electromagnetism: We introduce a series of individual pointlike test particles into the void-field and watch how each behaves. This is, in principle, no different from Galileo's falling-body experiments, described in chapter 5, which measured the properties of Earth's invisible gravitational field. The void is no longer empty with the presence of the test particle, but we can at least make a working assumption that we can treat it as a separate entity.[8]

Let me try to state, as succinctly as I can, what I am doing here—lest the reader think it is too weird. We have some void over there. We put in some test particles and measure their positions, velocities, and accelerations as a function of time with a meter stick and clock. From these measurements, we try to infer some properties that describe the void. And we will compare those properties with those of our nonempty universe.

We start by attempting to measure the position of our test particle at various times. However, we have no reference point with respect to which we can do this. The particle can have any position we choose. As shown in figure 8.1, we can "translate" the origin of the spatial coordinate axes—the reference point that we use to measure position—to any place we choose. In other words, all "places" are arbitrary. This leads us to our first observed property of the void: It contains no special position. We call this property *space translation symmetry.*

Space translation symmetry: There is no special position.

The next thing we notice is that we can walk around our test particle and view it from any angle without seeing any differences. As also shown in figure 8.1, we can rotate the axes of the spatial coordinate system to any direction we choose. In other words, all directions are arbitrary. This leads us to our second observed prop-

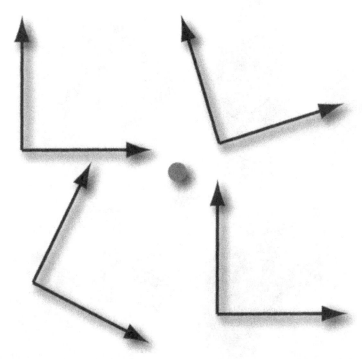

Fig. 8.1. Space translation and rotation symmetries. The position of a particle can be measured with respect to any coordinate system with arbitrary location of its origin and arbitrary direction of its axes.

erty of the void: It contains no special direction. We call this property *space rotation symmetry*.

Space rotation symmetry: There is no special direction.

Like our point particle, a perfect sphere has space rotation symmetry. No matter what angle you view it from, it looks the same. This symmetry is about any rotation axis you choose. By contrast, a cylinder has rotation symmetry about a single axis, the axis of the cylinder. A cone has similar symmetry. A snowflake also has symmetry about a single axis, but this symmetry is further restricted. While the cylinder or cone can be rotated by any angle about its axis and still look the same, the snowflake can only be rotated about angles in steps of sixty degrees. It has *discrete* symmetry, while the cylinder, cone, and sphere have *continuous* sym-

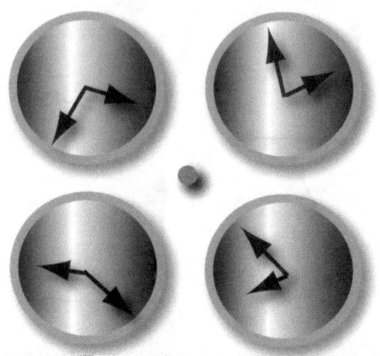

Fig. 8.2. Time translation symmetry. Time can be measured with respect to any arbitrary reference time.

metries. Note that as you move from a sphere to a cylinder to a snowflake, you are moving from a simple object to a more complex one, from less structure to more structure. Thus, complexity is associated with low levels of symmetry—what we call *broken symmetry.* In any case, the void has the highest level of spatial symmetry. And that's not nothing.

Now, what about time? We have a clock to measure time, but nothing is available to specify when we should start the clock, what time we should label as $t = 0$, the origin of our time measurements. Obviously any arbitrary point in time will do. That is, we can "translate" the reference time that we use to measure time intervals to any time of our choosing (see figure 8.2). In other words, all "times" are arbitrary. We call this property *time translation symmetry.*

Time translation symmetry: There is no special moment in time.

Thus, the void is characterized by certain properties that we describe in terms of symmetry principles. This is not to imply necessarily that these principles exist as some kind of Platonic reality of which the void is a manifestation, although that is one possible metaphysical interpretation. Here we are just acting like experimentalists who give no thought to metaphysics. We are trying to make measurements with meter sticks and clocks. Time and space are simply operational concepts we ourselves have invented. Time is what we measure on a clock. Distance is what we measure with a meter stick. So far, we have not been able to make such measurements and that fact, in itself, has revealed certain properties of the void. In terms of concepts that we have already used with great success in the description of phenomena in our nonempty universe, an empty universe must be described in a way that makes no assumptions about a special position and direction in space, or moment in time.

The concepts described so far should be easy to understand. At the same time, they are very profound. What makes them profound is the fact that they apply to our nonempty universe as well. Indeed, space translation symmetry was the great principle discovered by Copernicus, although, of course, it was not originally expressed in that fashion. Until Copernicus, Earth generally was regarded as a very special point in space—the fixed center of the universe (an assumption reflected in the Bible). As I have already noted, the Copernican revolution was a major paradigm shift that triggered the scientific advances which followed. It can even be argued that the growth of science began when the weight of this ancient misconception was cast aside, when humanity became humble enough to realize that it was not at the focus of existence.

SYMMETRY AND CONSERVATION

While students in elementary physics classes are not burdened with such abstract concepts as symmetry principles, these principles have become an integral part of modern physics. They lie in the background of all that is taught in physics classes and used in many practical applications—in engineering and other fields. The elementary physics student is trained to apply the laws of physics in practical problem solving, rather than to understand them in any deep way. The most important and useful of these laws are the great conservation principles of energy, linear momentum, and angular momentum. These are universal or global laws, that is, they seem to apply for all types of phenomena—from machines to chemical reactions to biological processes to stars and galaxies. As far as we can tell from astronomical observations, they apply everywhere in the universe and have done so for all time. Again, for clarity's sake, these laws are as follows:

Conservation of energy: The total energy of an isolated system is constant.

Conservation of linear momentum: The total linear momentum of an isolated system is constant.

Conservation of angular momentum: The total angular momentum of an isolated system is constant.[9]

A good part of the elementary physics student's time is spent solving problems by utilizing these powerful laws. Energy conservation, for example, enables one to compute the speed at which water will emerge from a small hole near the bottom of a tank of water, knowing only the height of the tank. Linear momentum conservation can be used to compute how fast a defensive back must run to stop a fullback in his tracks. Being less massive than the fullback, he has to run faster to cancel the fullback's linear momentum. Angular momentum conservation will tell you how fast a figure skater will spin when she pulls in her arms and explains why she speeds up.

In advanced, graduate-level physics courses, the student begins to learn of the deep connection between symmetry principles and universal conservation laws—that they are simply different ways of saying the same thing. Conservation of energy is shown to

Fig. 8.3. The connection between linear momentum conservation and space translation symmetry. The absolute position of an object of fixed momentum cannot be measured.

follow from time translation symmetry. Conservation of linear momentum follows from space translation symmetry. Conservation of angular momentum follows from space rotation symmetry.

Let me give just one example, for the case of linear momentum, that illustrates this connection. Consider an automobile moving along a highway at constant velocity, as shown in figure 8.3. Its mass is constant, and so, then, is its linear momentum, which is the product of mass and velocity. Suppose equally spaced, identical telephone poles are situated along the highway, which are used to mark the car's position at the time it passes that point. If we were to watch a film of the trip, we could not distinguish one pole, that is one position-marker, from another; that is, we have space translation symmetry.

Contrast this with the situation when the car brakes to a halt and its linear momentum is no longer constant. Watching the film in this case, we can single out one telephone pole as special—the one nearest where the car stops. Thus we see that a connection exists between conservation of linear momentum and our ability to single out a special position in space.

By a more sophisticated, mathematical analysis, the conservation laws can be proven to automatically exist for any physical system that possesses the corresponding symmetries and vice versa. This is called *Noether's theorem*, named after mathematician Emmy Noether, who proved it in 1918. These laws are true in classical mechanics and carry over to quantum mechanics. As we have seen, these are precisely the space-time symmetries of the void. The fact that phenomena in our nonempty universe obey these conservation laws, when the caveat of an isolated system is applied, indicates that our universe possesses the symmetries of the void—exactly what would be expected had it spontaneously appeared from an initial state characterized as the void.

So if someone asks where the laws of conservation of energy, momentum, and angular momentum came from, the answer is: They came from nothing. No act of creation was necessary to produce those laws. If these conservation laws were not found to hold in the universe, then we would be able to make a good case for a creator because that would indeed be unnatural, if not supernatural. But the conservation laws are found to hold. They are natural and provide a good argument for the absence of a creator.

RELATIVITY AND THE LAWS OF MOTION

Let us get back to our experiments on the properties of the void. Recall that we could define no special, or "absolute," position, direction, or moment in time. But if we cannot do this, how can we define the motion of the test particle? That requires a measurement of quantities such as velocity and acceleration with meter sticks and clocks. Velocity is the time rate of change of position, and acceleration is the time rate of change of velocity.

Thus, when Galileo championed Copernicus's idea that Earth moves about the Sun, contrary to scripture, he was challenged by some Church fathers with a very reasonable question: If Earth is moving, why don't we notice it? Instead of doubting his own telescopic observations, which convinced him of the reality of the Copernican view, Galileo attempted to make that view consistent with experience. He proposed that all *uniform* motion is relative, where uniform means constant velocity—motion in a straight line at constant speed.[10] Just as no special "frame of reference" exists to define an absolute position and direction in space, no special frame of reference exists that is absolutely at rest. We have no way of knowing whether a body is at rest or in uniform motion. This property is called the *principle of relativity*.

Fig. 8.4. The principle of Galilean relativity. An observer inside a closed capsule cannot measure the velocity of the capsule.

Principle of relativity: There is no absolute uniform motion.

One way to think of the principle is from the point of view—the *frame of reference*— of an observer inside a closed capsule in outer space (see figure 8.4). As long as the observer cannot see any outside reference point, such as Earth or the Sun, he has no way of knowing whether he is moving at constant velocity or is at rest. This would also be the condition for an observer all alone in the void. In that case, we do not even need a closed capsule, since there would be nothing outside to see. Today this principle is demonstrated every time we fly in a jet airliner. Except for vibrations from the engines, and occasional bumpy air, we have no sensation of hurtling though the sky at hundreds of kilometers an hour.

Similarly, but without the shared experience of jet airliner travel, Galileo pointed out, we stand on Earth with no sensation of hurtling through space at 30 kilometers per second. He explained that what we experience as "motion" is not velocity but *changes* in velocity—acceleration. Of course, the motion of Earth is not at constant velocity in its near-circular orbit around the Sun, but we do not sense this slight acceleration. Nor do we notice the acceleration that results from Earth's rotation about its axis, although, as we recall from chapter 5, that can be measured and is the source of cyclonic storms in the atmosphere. Again referring to our own experience, when we take off in an airliner, we are pressed against the back of our seats; when we land, we are pressed forward against our seat belts. What we experience is the acceleration and deceleration of the aircraft.

The principle of relativity is the first principle an elementary physics student learns because it constitutes the jumping-off point for the mechanics developed by Newton, who was born the year Galileo died. Newtonian mechanics is based on the three laws of motion.

Newton's Laws of Motion

(1) A body at rest will stay at rest, and a body in motion will stay in motion in a straight line with constant velocity unless acted on by an external force.

(2) The total force on a body is equal to the time rate of change of the momentum of the body.

(3) For every action there is an equal and opposite reaction.

Common experience tells us that a body at rest will stay at rest unless it is acted on by a force. But since, according to Galilean relativity, one cannot distinguish between a body at rest and a body in uniform motion, it follows that a body in uniform motion will stay in uniform motion with the same velocity unless acted on by an outside force. (Recall that the principle of relativity only excludes *absolute* uniform motion.) And Newton's first law of motion is exactly what would be expected for our test particle in the void. Later, in university-level introductory courses, students are shown that laws (1) and (3) follow from conservation of linear momentum and that (2) simply represents the quantitative definition of force as the agent for changing momentum. An isolated body has no forces acting on it and so has constant linear momentum. Since the linear momentum of a body is the product of its mass and velocity, when the mass is constant, the velocity remains unchanged in magnitude and direction.[11] The third law of motion (3) is also the result of conservation of linear momentum. When you fire a gun, you experience a recoil in which your backward linear momentum balances the forward linear momentum of the bullet, the total linear momentum thus remaining constant.

In 1905 Einstein's special theory of relativity extended Galileo's principle of relativity to the case where the relative motion of bodies was a large fraction of the speed of light. To make it agree with observations in electrodynamics,[12] Einstein found it was necessary to discard the common perceptions that the time interval between events and the distance interval between two points in space are absolute. By absolute, I mean the same in all reference frames. We have already seen that no absolute position in space or moment in time exists. Einstein showed that the same is true for distance and time *intervals*. These depend on the relative motion of the observer making the measurements. Einstein found that the speed of light in a vacuum, on the other hand, *was* an absolute constant that had to be the same for all observers regardless of their relative motion.

If you and I are in separate spaceships moving relative to one another and we observe two supernova explosions, we will in general obtain different measurements for the distance between the stars that exploded and the time interval between the two events. One of us might observe them at the same time, but the other will not. As Einstein pointed out, we cannot even properly speak of things happening simultaneously as reported by any two observers in relative motion. Similarly, we cannot speak of two events occurring at the same place in all reference frames.

The procedure that enables one to compare the space and time intervals between events observed in different frames of reference is called the *Lorentz transformation*, named after physicist Hendrik

Lorentz, who proposed it in the nineteenth century but did not realize its full implications.

We have already seen how rotational symmetry in familiar three-dimensional space leads to conservation of angular momentum. As proposed by Hermann Minkowski shortly after Einstein published the special theory of relativity, time can be mathematically treated as an added dimension in a four-dimensional space-time. If we make the assumption of rotational symmetry in four-dimensional space-time, then the Lorentz transformation follows. Thus the "laws" of special relativity, like the conservation laws, are also the natural consequence of a symmetry principle—*space-time rotation symmetry.*

Space-time rotation symmetry: There is no absolute direction in space-time.

When we combine all the translational and rotational space-time symmetries we have considered, we have what is called *Poincaré symmetry.* In our series of experiments with the void, we find that it possesses this set of symmetries. Poincaré symmetry is a property of the void, and apparently a property of our universe as well. Thus the principles of special relativity, as well as the great conservation principles, follow from and are consistent with the natural symmetries of the void. So far, we have found no need for a creator, a law-giver of any sort, in some of the most important "laws" of physics.

GRAVITY FROM THE VOID

In our experiment with the void, we found that we could define no special frame of reference in which a body was at rest. This agreed with Galileo's principle of relativity, extended by Einstein in the special theory of relativity. These principles applied to reference frames moving at constant velocities with respect to one another. However, for each of our test particles in the void, we also have no way of measuring its *acceleration*. Thus, at least as far as the void is concerned, whether or not a body is accelerating is also indeterminate.

Newton had made a distinction between accelerated and nonaccelerated reference frames. Newton's laws of motion seem to apply only to nonaccelerated systems, what we call *inertial* reference frames. A rotating system, like Earth, is a *noninertial* reference frame, and Newton's laws of motion can be maintained in such frames only by introducing "fictitious" forces such as the centrifugal and Coriolis forces. This has always puzzled physicists and philosophers. Suppose Earth were the only body in the universe, like our test particle in the void. How could we tell that it was rotating? What would it be rotating with respect to? Newton thought that it rotated with respect to absolute space. According to this view, space is something substantial.

Leibniz, however, rejected Newton's notion of absolute space, arguing that the concept of space was not separable from the relations of material bodies. If one has no matter, then one has no space—or time. This issue was intensely debated by some of the greatest philosophers for many years, with no resolution. In the nineteenth century, some progress was finally made when physicist and philosopher Ernst Mach proposed that inertia is the net effect of the mass of Earth and other celestial bodies. This is called *Mach's principle*. It implies that one cannot detect acceleration, such as rotation, in the void and that we can extend the principle of relativity to include all motion, not just uniform motion.

In 1907 Einstein introduced the *principle of equivalence between inertia and gravitation* that eventually led to his theory of gravitation called the *general theory of relativity*, which appeared in 1916.

Principle of equivalence: Gravitation and acceleration are indistinguishable.

Einstein used the example of a man inside an elevator in free fall (see figure 8.5). Every object inside the elevator will fall together, and the unfortunate person inside would have no way of knowing he was falling and not simply safely floating in a capsule in empty space.

Fig. 8.5. The principle of equivalence. An observer inside a falling elevator cannot determine if he, in fact, is falling in a uniform gravitational field or is in outer space far from any gravitational fields.

The principle of equivalence can be formulated in a way similar to what I did above for the principle of Galilean relativity. Recall that this form of relativity says that an observer in a closed capsule cannot distinguish between being at rest and being in motion at constant velocity. The principle of equivalence states that an observer in a closed capsule cannot distinguish between being accelerated and being in the presence of a uniform gravitational field. The caveat "uniform" is added here since it is in practice possible to tell you are on Earth rather than in an accelerated capsule by noting that objects dropped from different places will not fall along parallel lines but along lines that converge to the center of Earth.

Einstein proceeded to write down the equations that mathematically describe this situation. These equations had to conform to what the falling observer would see, namely, no evidence for a gravitational field. And, when properly transformed to the frame of reference of an observer on Earth, the equations had to conform to what the Earth-bound observer would see, namely, bodies accelerating measurably toward Earth. When Einstein did this, he obtained the equations of general relativity.[13]

Thus, in the modern view, gravity does not exist as a physical force field that surrounds a massive body, a tension in some elastic substance, such as the aether, as once was thought to be the case. Rather, gravity is an artifact—a fictitious force—that we introduce into our equations to describe what happens when two bodies accelerate toward each other with no evident interaction between them and no otherwise detectable "field." Those equations must reflect the fact that an observer in a closed capsule in free fall cannot detect the fact that he is accelerating. This is precisely the situation that would occur if the capsule were in an otherwise empty void. In that case, acceleration cannot be defined. General relativity is just the extension of the principle of relativity to accelerated reference frames.

I may have oversimplified the case. In order to describe gravity as a fictitious force, Einstein found it was necessary to discard the usual description of space in terms of *Euclidean* geometry. Long before, mathematicians had shown that one can relax Euclid's geometric axiom that parallel lines never meet. Einstein was the first to find a physical application for the resulting non-Euclidean geometry, using it to describe four-dimensional space-time in general relativity.

An example of a non-Euclidean space is the two-dimensional

surface of a sphere. The lines of longitude on Earth are parallel, and yet they meet at the poles. On the surface of a sphere, the path of a great circle (any circle whose center is at the center of the sphere) is called a *geodesic*. In general, in any given geometry, the geodesic is the shortest distance between two points and is a straight line only in Euclidean space.

In classical mechanics, free particles—particles not acted on by any force—follow a geodesic path in space. Usually we think of this as a straight line, which is a good approximation until we get to the astronomical scale. In general relativity, bodies still follow geodesic paths, but these paths can be curved. The curvature of the geodesic a body follows is determined by the masses of other bodies in the vicinity. Thus Earth travels in a geodesic around the Sun. That motion is "more natural" than a straight line because the presence of the Sun "warps" the space around it.

While this might sound very complicated, and the mathematics takes some training to learn, the equations describing these geodesics follow from the assumption that they are the same—the technical term is *invariant*—from one frame of reference to the next. That is, they follow from a generalized principle of relativity that maintains the symmetries of the void.

Thus, once again, we have a great scientific theory, general relativity, following from properties of the void. This concept—that much of physics is a formalized description of the properties of the void—is very profound. It is also very simple. Indeed, it is profound by virtue of its simplicity. The most powerful and universal principles that have been discovered by physicists turn out to be exactly those which result from taking great care to assure that their mathematical descriptions of phenomena possess the symmetries of the void. These great principles could not be otherwise, if the universe exploded, uncreated, out of the void. Thus, not only do these laws provide no evidence for a creator, they constitute a strong case for his nonexistence.

QUANTUM MECHANICS

Quantum mechanics is usually pictured as a great revolutionary break, a giant "paradigm shift," away from classical mechanics. While many of its philosophical implications, especially indeterminism, were perhaps revolutionary, the actual theory as originally developed independently by Heisenberg and Schrödinger follows rather directly from classical mechanics. Heisenberg made basically only one innovation: he postulated that observables be represented in the theory by noncommuting operators, that is, mathematical objects that do not obey the commutative law of real numbers, $AB = BA$. In the remainder of his program, he used the classical equations of motion as his guide.

Similarly, Schrödinger inferred his famous wave equation from the classical Hamilton-Jacobi equation of motion, which has the form of a wave equation. I have already mentioned that Noether's theorem applies to quantum mechanics as well as mathematics. While the mathematical derivation is beyond the scope of this book, it can be shown that quantum mechanics also follows in a natural way from classical mechanics.

FORCES, STRUCTURE, AND
SPONTANEOUSLY BROKEN SYMMETRIES

We have seen how much of physics—conservation of energy, linear momentum, angular momentum, and the principles of special relativity and general relativity—are properties of the void. We have seen that our universe possesses these properties, strongly suggesting it evolved naturally from the void. This has brought us a long way, to an understanding of the origin of the most important "laws" of physics. However, we have still not accounted for the source of order and structure we see in our universe. This still leaves a gap for a god to be situated, and, indeed, this is where some physicist-theologians think they see an opening to enable them to reconcile their faith with science. Nevertheless, we can continue to refute the need to invoke divine design to explain the laws of physics by providing an alternate scenario grounded in natural processes.

Let us begin by going back to Newton's second law of motion. When a body is not isolated, the law of conservation of momentum does not apply, and its linear momentum can change. The total force on the body is then quantitatively defined, by Newton's second law of motion, as the time rate of change of the linear momentum. A similar set of laws exists for rotational motion, and they relate in the same way to angular momentum. *Torque* is the time rate of change of angular momentum. A rotating rigid body will remain rotating at constant angular velocity unless acted on by an external torque. If the body is not rigid, as in the case of a figure skater doing a spin, the angular velocity can change, but the angular momentum still remains constant.

We can see how a force can come about in our experiments with the void by introducing a second test particle. Now we can begin to use our meter stick to measure the distance between particles, and we are no longer free to translate the position of one particle independent of the other, although we can still translate the two together. Thus we can expect linear momentum conservation to apply to the system composed of the two particles, which is still isolated, but not to individual particles themselves. They can interact with one another, exchanging momentum. The simplest way to do that is for particle A to emit a third particle, X, which carries momentum across to particle B where X is absorbed. The total momentum and energy is conserved, but particles A and B recoil away from one another. Indeed, this is precisely the picture used in the standard model of elementary particles and forces.

Since conservation of linear momentum follows from space translation symmetry, the action of a force has the effect of breaking that symmetry. Or you can think of the act of breaking the symmetry as producing the force. (Note how causal language creates trouble here, unless you assume cause and effect are interchangeable, which they are at the fundamental level, as I show in *Timeless Reality*). In either case, it then becomes possible to single out one point in space as different from all others. For example, a ball tossed in the air will experience a change in direction when it reaches its highest point and then starts to come down. That point is distinguishable from all the others along the path of the ball. Similarly, the action of a torque is to break rotation symmetry, selecting out a special direction of space, namely, the axis around which the torque is applied.

Forces and torques, then, are associated with broken symmetries. These broken symmetries supply complexity to the universe, as when water vapor freezes to produce a snowflake with less symmetry and, with the generation of information, greater complexity. The connection between forces and broken symmetries was one of the great discoveries of twentieth-century physics and provided the key for the development of the highly successful standard model of particles and forces. In the standard model, the electromagnetic, strong nuclear, and weak nuclear forces are all described in terms of spontaneous broken symmetries. Here the term "spontaneous" refers to the fact that the symmetries are not broken by some external deterministic action, by neither natural nor supernatural law, but randomly.

The basic equations of the quantum field theory comprising the standard model contain underlying symmetries—not just those of space and time already discussed but symmetries that relate to other "inner" dimensions. These symmetries are assumed to apply at extremely high energies, or temperatures, as existed in the first tiny fraction of a second of the big bang. All the particles and forces would have been originally identical, with equal amounts of matter and antimatter. As the universe expanded and cooled, these symmetries, according to the model, were spontaneously broken, leading to differences among the particles and forces and a large excess of matter over antimatter. In this way, the structure of the universe may have appeared, without cause or creation, the way a snowflake appears out of a less structured and more symmetric cloud of water vapor.

THE INSTABILITY OF THE VOID

Now we are finally in a position to begin to shed a little light on the common question: Why is there something rather than nothing? The unstated assumption of this question is that nothing is somehow a more natural, stable state of affairs than something, and so we must explain the existence of something. If, on the other hand, something is more natural and stable than nothing, then the correct question is: Why should there be nothing rather than something?

Now, as I have already indicated, I am at a loss to describe "nothing" outside of a well-specified ontology such as atoms and the void. If the term refers to "no thing," then it has no properties to describe.[14] It serves no useful purpose to rehash here the ancient philosophical debate over the meaning of the words "something" and "nothing." Therefore, I have chosen to stick to what I know best, empirical physics in which quantitative measurements are made with specified instruments, and theoretical physics in which those measurements are described mathematically. I accept the fact that all measurements are "theory-laden," in the sense that theory is used to define the terms of the measurements. I understand the process is not totally objective, but neither is it totally subjective. Rather, the procedure is "intersubjective." Common agreement is reached by multiple individuals on the terms by which the measurements will be performed, and these measurements are then carried out by multiple, disinterested, or even skeptical observers. Only when the observers agree on the results of a certain set of measurements are they taken seriously.

I have introduced the notion of the void as a physical concept—an emptiness that still can be studied by the introduction of test particles whose relative positions and movements can be measured with meter sticks and clocks. This does not mean that the void is a real substance or object. It does not kick back when we kick it. However, the test particles kick back when we kick them.

We have seen that our nonempty universe possesses space-time properties which coincide with those of the void, strongly indicating that it could have formed from the void. We have seen that those properties can be expressed in terms of familiar principles of physics. We have seen that such a transition, from emptiness to nonemptiness, violates none of these principles of physics. But if the universe was a void, then why did it not stay a void? What nudged it from emptiness to nonemptiness? Maybe God had to do it after all!

Common sense would seem to say that the entity we call the void is the more natural state of the universe. Theists continually argue that the transition void-to-universe must have been a miracle because of the decrease in entropy, or increase in order, that they imagine must have occurred, in apparent violation of the second law of thermodynamics. However, as we saw in chapter 6, the initial state of the universe was one of maximum entropy, the same as a formless void. Thus no violation of the second law occurred. Indeed, this is yet another property that the universe shared with the void, at least at that first moment—maximum entropy.

If we have learned anything at all from the past four hundred years of discovery, it is that common sense often leads one astray. It is common sense that Earth is flat and immovable at the center of the universe. It is common sense that we can always tell when we are in motion. It is common sense that time and space intervals are the same for all observers. It is common sense that living species and shining stars are fixed and immutable. All these, and many other conclusions based on common sense, have been shown to be wrong.

In fact, the physical world is filled with examples in which the more complex system is the more stable. I have used the snowflake several times as an example, and it can be used again here. The so-called phase transitions that familiar matter undergoes in moving from the vapor state to the liquid or solid states are changes from states of lower to higher complexity. Heat from the Sun is required to take something complex, like solid matter, and make it into something simpler, namely liquid or vapor. Take away the Sun and all the vapor and liquid on Earth will freeze to solid matter, as they have on distant Pluto. The more complex solid state of matter often is the more stable one.

The Snowflake

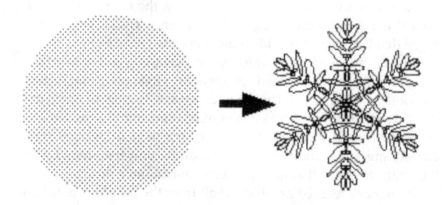

Sphere of water vapor

Snowflake

Fig. 8.6. The production of order by spontaneous symmetry breaking is exemplified by the snowflake. The snowflake has less symmetry but more order than the original sphere of water vapor out of which it formed as the temperature dropped below freezing.

Note that these phase transitions do not violate the second law of thermodynamics. When a snowflake forms, as shown in figure 8.6, its loss of entropy is compensated by a gain in entropy of the surrounding air. If we could take that snowflake and keep it isolated from the rest of the universe, it would forever remain a snowflake. Heat is necessary to melt it back to its simpler state.

Another example is an iron magnet or ferromagnet, shown in

The Ferromagnet

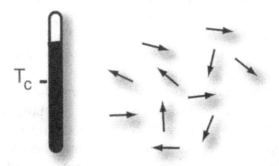

No magnetic field at high temperature

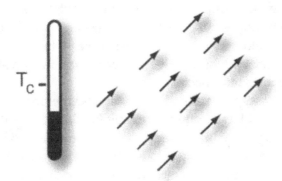

Magnetic field appears when temperature drops below critical value. Its direction is random

Fig. 8.7. The production of order by spontaneous symmetry breaking as exemplified by the ferromagnet. The ferromagnet has less symmetry but more order than the original randomly oriented magnetic domains out of which it formed as the temperature was reduced below the Curie point.

figure 8.7. The stable state of a ferromagnet is one in which the magnetic domains inside an iron bar line up, giving a net magnetic field. To destroy the magnetic field, you must heat the magnet above a certain critical temperature, called the Curie point, so that the domains move around randomly, canceling out each other's field. Note that the nonmagnetic state is simpler, more symmetric—and less stable—than the magnetic state. That is, the natural state of the iron bar is the more complex one, the one with broken symmetry.

Now, ferromagnetism is well described in terms of the theory of electromagnetism that was consolidated into a few equations in the late nineteenth century by Hendrik Lorentz and James Clerk Maxwell. Microscopic quantum effects in electromagnetism are fully accounted for in the theory of quantum electrodynamics, developed in the 1950s, that is now part of the more general standard model which also takes into account the nuclear forces. The basic equations of electromagnetism are symmetric to rotations about any axis. The ferromagnet, which singles out a particular direction—the direction of its net magnetic field—thus spontaneously breaks the symmetry of the underlying theory. Nothing in the theory says this breaking must take place along some specific axis.

If one starts with a spherical ball of iron at a temperature above the Curie point, it will have no net magnetic field, and thus it has rotational symmetry. The symmetry is spontaneously broken when the iron ball cools below the Curie point; but the direction of any net magnetic field that results is unknown, indeed undetermined, ahead of time. The direction that results is random. If one had a large number of balls of iron to begin with, a random distribution of directions would result and, overall, rotational symmetry would be maintained statistically. Thus the breaking of rotational symmetry by any given iron ball is "local," while rotational symmetry remains in force "globally."

As I have mentioned, spontaneously broken symmetry was one of the key concepts that led to the development of the current standard model of elementary particles and forces. In this picture, empty space—the void—is not stable. For this reason, it is referred to as the *false vacuum*. We can think of it as analogous to a hot ball of iron with no magnetic field. The lower energy *true vacuum*, on the other hand, is like a cool iron ball in a more stable state in which a magnetic field is present. In this case, it is not rotational symmetry that is broken, and the field is not magnetic, so the analogy is not exact. But still, the nonzero field has an energy density and negative pressure.[15] As we have seen, this is precisely the condition that can lead to the exponential expansion of the universe that we call inflation.

Consequently, in the scenario suggested by modern physics and cosmology, the simple, symmetric void empty of matter transforms into the complex, asymmetric universe filled with matter because that universe is more stable than the void. We have a universe rather than a void because it is more natural. As the old saw goes, nature abhors a vacuum. If we had only emptiness, then we would have to ask why there was emptiness rather than fullness. But then, we would not be here to ask the question.

NOTES

1. Hugh Everett III, "'Relative State' Formulation of Quantum Mechanics," *Reviews of Modern Physics* 29 (1957): 454–62; Bryce DeWitt and Neill Graham, eds., *The Many-Worlds Interpretation of Quantum Mechanics* (Princeton: Princeton University Press, 1973); David Deutsch, *The Fabric of Reality* (New York: Allen Lane, 1997).

2. For more discussion, see my *The Unconscious Quantum: Metaphysics in Modern Physics and Cosmology* (Amherst, N.Y.: Prometheus Books, 1995) and *Timeless Reality: Symmetry, Simplicity, and Multiple Universes* (Amherst, N.Y.: Prometheus Books, 2000).

3. Paul Davies, *The Mind of God: The Scientific Basis for a Rational World* (New York: Simon and Schuster, 1992).

4. See Carl Sagan's introduction to Stephen Hawking's best-seller, *The Brief History of Time: From the Big Bang to Black Holes* (New York: Bantam, 1988).

5. For a popular exposition, see Brian Greene, *The Elegant Universe: Superstrings, Hidden Dimensions, and the Quest for the Ultimate Theory* (New York: W. W. Norton, 1999).

6. Thanks to Brent Meeker for suggesting that I put it this way.

7. Momentum and energy are, in fact, the Fourier-transformed variables of spatial coordinates and time. So they are equivalent.

8. Some physicists might argue that such separation is impossible, in the light of the "nonlocality" of quantum mechanics. But in *Timeless Reality* I showed that this kind of separation can be done if one allows for time reversibility.

9. Here I have made the distinction between linear momentum and angular momentum that is familiar in elementary physics. The first is what is often simply called momentum, the product of a body's mass and its velocity. Angular momentum is defined with respect to some axis and is the linear momentum times the perpendicular distance to that axis. Also, for simplicity, I have not stated these laws in the most general way. Under certain conditions, one or two of the conservation laws will apply for nonisolated systems.

10. In physics we make a technical distinction between speed and velocity. Velocity is a vector that has magnitude and direction. Speed is the magnitude of velocity.

11. If the mass is not constant, as for a rocket, then the velocity can change, but the momentum will remain constant.

12. By this time, electromagnetism was well described by a set of equations developed by James Clerk Maxwell. Einstein sought to explain the already well known fact that these equations violated Galilean relativity.

13. For a good discussion of this and further references, see Steven Weinberg, *Gravitation and Cosmology: Principles and Applications of the Theory of Relativity* (New York: John Wiley, 1972). While this book is highly technical, with many equations, it is still possible to read between the equations and obtain some valuable insights, since Weinberg is an excellent writer as well as a brilliant physicist.

14. For recent discussions of "nothing," see John D. Barrow, *The Book of Nothing: Vacuums, Voids, and the Latest Ideas about the Origins of the Universe* (New York: Pantheon Books, 2000) and Robert Kaplan, *The Nothing That Is: The Natural History of Zero* (Oxford: Oxford University Press, 1999).

15. In *Timeless Reality*, I argued that all fields are simply mathematical ways we describe a more fundamental reality of particles, which exist in the theory as the "quanta" associated with the fields. This does not affect the current discussion, in which it can be specified that I am simply using the field description and not implying any deep, metaphysical reality to these fields.

NINE

ABSENCE OF
EVIDENCE

They are looking in utter darkness for that which has no existence whatsoever.
—Desiderius Erasmus, 1509

NOT REQUIRED TO EXIST

S o far, we have only talked about the various arguments for the
existence of God that are based on the common intuition that
the universe, as we know it, could not have come about and have
the structure it has except by the action of some transcendent
reality. We saw that this intuition is expressed in a number of ways,
but they all amount to little more than variations on the ancient
argument from design. While this argument, as traditionally presented, is well known to be logically flawed, it has reappeared in
recent years expressed in the language of science rather than that

of philosophy or theology. And it is in scientific language that I have addressed the issue.

I have attempted to show that current scientific knowledge provides a plausible scenario by which the universe, as we observe it, can be self-contained—not requiring external creative actions to either come into existence or to maintain that existence. This scenario cannot be proven and may have few, if any, consequences that can be tested by observations. Still, as long as we limit ourselves to established scientific knowledge and discuss its implications, such a discussion remains within the bounds of science.

Those observations we have made, and those theories we have successfully tested, carry with them a great store of credibility just by virtue of their acceptance into the ranks of scientific knowledge. They can be used to make extrapolations into realms beyond the observable and testable. Of course, we must admit the possibility that these interpolations and extrapolations will turn out wrong. However, this fact does not require that we avoid mentioning them altogether. They are informed speculations, the best we can do, and still better than speculating from a basis of no knowledge whatsoever, or worse, abandoning the quest.

The scenario I have presented is consistent with all scientific observations and is based on scientific theories that are well tested. It is proposed with the modest purpose of providing a counter-example that serves to refute any assertion that the universe and its laws, as revealed by scientific data, cannot be explained without introducing an external, supernatural creator. Again, I do not claim that this proves that such a creator does not exist, only that one is not required to exist according to our best current knowledge. This makes belief in such a creator optional, but an uneconomical option with no basis in science.

In this chapter we move to a discussion of the down-to-earth observations that are frequently introduced to profess direct evidence for the existence of a deity who plays an active, everyday role in human life. I will show that we can find no such evidence. And, yet again, because I can predict the line of criticism that this book will generate, I need to make it clear up front that I am not claiming that the absence of evidence eliminates all possibilities for a god to exist in every conceivable form. And, I am not reviewing all the theological and philosophical arguments for or against God. I am simply evaluating the *scientific* arguments and *scientific* evidence claimed for a deity according to the same criteria that science applies to any extraordinary claim. I conclude that, so far, they fail to meet the test.

PREDICTION AND PROPHECY

Although the Bible should properly be viewed as a religious rather than a scientific document, believers often assert that it contains precise predictions, or *prophecies*, which have come to pass. These are then used to testify to the Bible's accuracy as a source of revelatory information about another, spiritual, plane of reality that transcends material reality.[1] I would not even raise the issue of the Bible in this present book of science were it not for the claims of successful predictions comparable to those of science. These ask to be evaluated on a scientific basis.

The ability of science to predict the future is an important quality that testifies to its power and provides a good basis for believing that it correctly deals with at least some aspect of objective reality. The successful prediction by Edmund Halley (d. 1742) that a comet would reappear in 1758 did much to convince people of the validity of Newtonian physics. In 1846 French astronomer Urbain Jean Joseph Leverrier (d. 1877) accounted for perturbations in the orbit of the planet Uranus by proposing the the existence and position of a new planet. That same year, German astronomer Johann Gottfried Galle discovered the planet, now called Neptune, within one degree of the predicted position. Einstein's 1915 theory of general relativity predicted the bending of light by the Sun. Observations made shortly afterward by Sir Arthur Eddington during a solar eclipse in Africa seemed to verify that prediction, however, these were ambiguous. Nevertheless, many increasingly accurate observations since have solidly confirmed the validity of Einstein's prediction.

In 1931 Wolfgang Pauli predicted the existence of an elementary particle with zero electric charge and zero or very low mass that Enrico Fermi named the *neutrino*. This particle was observed in the laboratory in 1956 by Frederick Reines and Clyde Cowan. Neutrinos have now been observed from the Sun and from a 1987 supernova explosion in one of the Milky Way's satellite galaxies, the Large Magellanic Cloud.[2] In the 1970s physicists used the new standard model to predict the existence of fundamental particles called the W- and Z-bosons, which were theorized to account for the weak nuclear forces. These forces provide for the energy of the Sun and make our very existence on Earth possible. The bosons were observed in experiments with exactly the masses and other properties predicted. And as a final example, I have already mentioned the successful prediction of the inflationary big bang model that the cosmic microwave background should have a temperature fluctuation of one part in 100,000 (thirty-millionths of a degree Celsius).

These predictions were not just vague generalities, such as "Two nations will go to war and one will win" (Woody Allen, in a takeoff on Nostradamus), or sure things, like "The Sun will rise tomorrow." The scientific predictions I have mentioned were all *risky*. In each of these examples, the failure of the prediction would have falsified some part of the theory. I can give examples of falsified theories as well, such as an attempted extension of the standard model whose prediction that protons would disintegrate at a certain rate was not verified by experiment, thus eliminating that particular theory.[3]

Early in the twentieth century, philosophers Rudolf Carnap and Karl Popper proposed that the falsifiability of a theory is what defines it as science. While this proposition is no longer regarded as either necessary or sufficient for a theory to be scientific, falsifiability remains a major component of scientific method. A theory that can be falsified by some empirical test carries much greater weight than one that only explains observations after the fact. A risky prediction of a theory is one that could falsify it, and when the theory passes such a test, it earns more credibility.

Examples of successful, risky, falsifiable predictions, and I have only mentioned my personal favorites among thousands, provide us with the confidence to view as reasonable those other assertions science makes that may not be immediately testable. They also provide good reason to believe that science deals with the real world, not one of myth and fantasy. This does not mean that all the assertions scientists make will necessarily prove correct, but rather that they should be taken more seriously than a simple, random shot in the dark. Similar success for the prophecies of the Bible, or any other sacred text, would equally well enhance the credibility of its many assertions that may similarly not be amenable to scientific test.

Most of the biblical predictions that theists point to can only be checked within the Bible itself, which makes them self-fulfilling. In particular, hundreds of "incredible" prophecies referred to by Christians have to do with the ultimate appearance of Jesus as the long-awaited Messiah. This prediction and many others can only be checked by referring back to the Bible for verification. Furthermore, these events are being interpreted today, thousands of years after they are supposed to have taken place. The prediction of the reappearance of Halley's comet in 1758 would have carried far less force had it been made by modern astronomers rather than one who died in 1742.

Obviously, an event cannot be said to be predicted when it occurred before the prediction itself was made. It is easy to predict that if we go to the library and pick up a book on recent American history, published in 1964 or later, we will read that President John F. Kennedy was assassinated in November 1963. Note that the prediction was that we would find the reference, not that JFK would be assassinated at some future date. To have found such a reference in a book published, say, in 1962 would have been remarkable (unless it was a book by Lee Harvey Oswald).

On some occasions, one may successfully predict the uncovering of a hitherto unknown fact about an event that occurred prior to its prediction. For example, an archaeologist might predict that ruins of a city destroyed by war will be found if one digs at a very specific site. This is still a prediction of a future event, namely the discovery of the ruins.

In the case of the Bible, one cannot point to a single risky prediction that came true based on evidence outside the Bible itself. Here are just a few of the prophecies of the Bible of specific events that failed to materialize:[4]

- Gen. 13:15, 15:18, 17:8, 28:13–14. God promises Abraham and his descendants all of the land of Canaan. Four thousand years later, this promise has still to come to pass.
- Isa. 17:1. Damascus is predicted to cease to be a city. In fact, Damascus is one of the oldest continuously inhabited cities.
- Jer. 49:33 predicts that Hazor will become an everlasting wasteland in which humans will never again dwell. The King James Bible says it will become inhabited by dragons. None of this has happened.
- Zech. 10:11. The Nile is predicted to dry up. This has not yet happened.
- Ezek. 29, 30. The land of Egypt will be laid waste by Nebuchadnezzar, all its people killed and rivers dried up. It will remain uninhabited for forty years. This did not happen.
- Matt. 16:28, 23:36, 24:34; Mark 9:1, 13:30; Luke 9:27. Jesus tells his followers that he will return and establish his kingdom within a generation, before the listeners die. This did not happen.

Apologists can always come up with explanations for the failures of biblical predictions. One, often heard, is that the deity keeps himself hidden so that humans can exercise their free choice in believing or not believing. I cannot refute that possibility, though I have no reason to believe it. The fact remains that one does not have good reason to look to any document as an accurate source of information when that document is filled with provable falsities.

At best, the errors of the Bible can be classified as ambiguous failures. More convincing would be unambiguous successes. For example, suppose, for the sake of argument, that Jesus had predicted that a human would walk on the moon two thousand years in the future. Now, wouldn't that have been something? If he or other biblical figures had made any successful, risky predictions of this type, by whatever means, then we would have a difficult time not taking their teachings seriously. As it stands, we have no basis for doing so.

I have already mentioned the statements in Genesis that imply predictions of scientific observations not yet made at its writing but long since proven false, in particular the immobility and immutability of Earth and the stars, and Earth existing before the Sun, Moon, and stars. There are also frequent references to the "ends of the Earth" and the "four corners" resting on pillars.[5] Now, of course, these mistakes are perfectly understandable if we assume that the writers were normal humans, since they reflect the knowledge of the time. But if the Bible is in fact the revealed word of God, then this would be more evident if it provided an accurate model of Earth and the cosmos.

If we examine the Bible as a scientific document, we would have to conclude that it has been long since falsified. Again, I do not say that it is a scientific document, and most believers do not take it as such. However, the highly visible conservative Christian media, in the United States at least, continually claim that the events described in the Bible have been seconded by science. I am responding to that claim. This is not to deny any value that the scriptures of all religions may have as literary or moral texts, nor their obvious impact on human history. But because of the lack of success in predicting the future, I suggest that they tell us far more about ourselves than about the nature of ultimate reality. And if science does not always provide an inspiring, poetic image of humanity that reinforces our feelings of self-worth, then at least it should be far more trustworthy than biblical religion in offering a glimpse of the reality that transcends petty human concerns. Perhaps science can teach us humility and self-reliance, and the need to live our lives within the universe as it really is, not as we might wish it to be.

STATISTICAL SIGNIFICANCE

Before proceeding, I need to discuss an important technical issue that will come up in this chapter and the next. When experimental results are presented in scientific papers, they are always accompanied by some estimate of their "statistical significance." While sophisticated statistical techniques are sometimes used, most often the authors simply present what statisticians call the *p-value*. For example, suppose an effect is reported with a *p*-value of 5 percent. This means that in a long sequence of identical experiments we would expect to observe an effect as great or greater produced by statistical fluctuations in 5 percent of the cases. This is not the same as "the likelihood [or probability] that the results were due to chance."

In a number of papers and books to which I refer below, the *p*-value is incorrectly characterized as the "probability against a reported effect being the result of chance." Sometimes "the odds against chance" are presented. Thus, a *p*-value of 5 percent is reported as the odds against chance being 20 to 1. These characterizations are wrong. In fact, it is always possible to get any observed effect by chance. You simply have to repeat the experiment enough times.

As we will see, scientific journals in various fields have differing standards for the threshold *p*-value they require for publication. In medical journals, this is typically 5 percent. This implies that on the order of every twentieth published experiment reports a result that is simply statistical fluctuation. In physics, by contrast, the threshold *p*-value is typically 0.01 percent, which means that only on the order of one in ten thousand published effects are simply chance artifacts.

I do not mean to make too serious an indictment of medical journals. Their purpose is to disseminate information to the medical community that may save lives. To serve that purpose, a certain amount of useless information that is simply the result of statistical artifacts may be acceptable. Still, allowing 5 percent useless information into the sample probably does more harm than good, as patients spend their money on worthless treatments and doctors are distracted from more effective courses of treatment.

In fields such as physics, no urgency exists in disseminating new results and a higher priority is placed on avoiding statistical artifacts. The physics standard takes into account the number of trials of various sorts that are made in typical physics experiments. The publishing threshold is sufficiently strict so that few statistical artifacts are published, but not so strict that nothing is published. The typical threshold p-value of 0.01 percent represents a thoughtful compromise that allows some noise in while making sure that clear signals are not missed. Ultimately, independent replication is still needed to confirm the result of a lone experiment.

THE BIBLE DECODED

Most scholarly discourse on the Bible bears on history and theology, not science, and so is beyond the scope of this book. However, one rather bizarre and widely reported recent incident involving the Bible does merit discussion within a scientific context. It offers an interesting, highly specific example of how one can be misled by relying on intuitive, commonsense estimates of the likelihood of chance events rather than sitting down and calculating them properly. This is a mistake that is easily made when one fails to count all the possible ways a result can happen by chance, a problem we have already seen with the cosmological fine-tuning and design arguments in previous chapters.

In 1994 Doron Witztum, Eliyahu Rips, and Yoav Rosenberg reported in the journal *Statistical Science* that they had discovered coded information in 78,064 characters of the Koren version of the Hebrew text of the *Book of Genesis*.[6] In this approach they were following an ancient Kabbalistic tradition of seeking messages hidden in the Torah. Rips is a prominent mathematician at the Hebrew University in Jerusalem. Witztum is an unaffiliated physicist, and Rosenberg is a computer programmer. Specifically, the names of a preselected set of famous medieval rabbis (as published in the *Encyclopedia of Great Men in Israel*) were found in close proximity to their birth or death dates, which the authors found surprising.

Witztum and collaborators used a procedure called *equidistant letter sequences* (ELS) in which one starts with a given letter in the text and repeatedly skips a fixed number of letters, ignoring spaces and punctuation. For example, if you take the first letter of the Book of Genesis, *T*, and skip every forty-nine letters, you will spell out TORH, the Hebrew spelling of Torah, the Jewish Bible. This was discovered fifty years ago by Rabbi H. M. D. Weissmandel, along with some other unexpected correlations. Applying ELS by hand is obviously tedious, but modern computers make it possible to try a vast number of characters with many different skip numbers and to seek out a large variety of correspondences. One finds that the word TORH appears 56,769 times in the Torah by this method.[7] Rips and his coauthors insisted that their results on the rabbis could not be the result of chance, claiming a statistical signficance of 1.6×10^{-5}.

In introducing the article, one of the editors of *Statistical Science*, Robert E. Kass, wrote: "Our referees were baffled: their prior beliefs made them think the Book of Genesis could not possibly contain meaningful references to modern-day individuals, yet when the authors carried out additional analyses and checks the effect persisted."[8]

Trying a similar analysis with 78,064 characters of the Hebrew version of the Book of Isaiah and a Hebrew translation of Leo Tolstoy's *War and Peace*, Rips and his coauthors said they found many names in close proximity to birth or death dates, but the results were statistically insignificant compared to Genesis.

In 1997, after spending time with Rips in Israel, journalist Michael Drosnin authored a popular book, called *The Bible Code*, which became an instant best-seller.[9] The Rips paper is reprinted in the book, making it perhaps the most reprinted scientific article of all time. Drosnin followed a decoding procedure similar to Rips and his collaborators, arranging the 304,805 Hebrew letters of the Bible into a large array, skipping punctuation and spaces. However, his method was not as carefully controlled. Instead of looking strictly for information specified beforehand, Drosnin seems to have searched thousands of letter sequences until he found something that looked interesting.

Drosnin writes that he discovered John F. Kennedy associated with "Dallas," and "Oswald" as "name of assassin who will assassinate." "Oswald" is also encoded with "marksman" and "sniper" as well as the description of the assassination: "he will strike in the head, death." The name "Ruby" appears connected with "he will kill the assassin." Drosnin is particularly fascinated with assassinations. In his most dramatic claim, the name of Israeli Prime Minister Yitzhak Rabin was found to be connected with "assassin will assassinate" prior to the actual event. Drosnin also leads the reader to think he determined that the assassination would happen in the Hebrew year beginning in late 1995, but this "prediction" was actually done after the fact.

On September 1, 1994, Drosnin flew to Israel with a letter reporting the code's prediction. It was given to Rabin, but on November 4, 1995, Rabin was assassinated in Tel Aviv. The assassin's name and the location of the tragic event were *later* discovered by Drosnin in the code.

In his book, Drosnin also asserted that the Bible predicted the assassinations of Anwar Sadat and Robert F. Kennedy. He uncovered mention of both World Wars, the Holocaust, Hiroshima, and the Moon landing: "Man on Moon," "spaceship," "Apollo 13," along with the precise date of the landing. The collision of the comet Shoemaker-Levy with Jupiter was also foretold. Drosnin found reference to the Watergate affair: "Watergate," "Who is he? President, but he was kicked out." "Shakespeare" appears with "presented on stage," "Macbeth," and "Hamlet." "Einstein" appears with "science" and "he overturned present reality." The Oklahoma bombing is encoded in detail, including "his name is Timothy," "McVeigh," "day 19," "on the 9th hour," "in the morning," "he ambushed, he pounced, terror," and "two years from the death of Koresh." No mention is made of the September 11, 2001, destruction of the World Trade Center, which happened after the book was published.

Well, you might ask, if this isn't evidence for the veracity of the Bible, what could be? How can science possibly explain these amazingly accurate prophecies of events that actually came to pass? Easy.

Retired U.S. Defense Department cryptologist Harold Gans had earlier examined the work of Witztum and his collaborators and found their methods sound. Drosnin reports this in his book. However, in June 1997, after *The Bible Code* was published, Gans issued a public statement that included the following:

> The book states that the codes in the Torah can be used to predict future events. This is absolutely unfounded. There is no scientific or mathematical basis for such a statement, and the reasoning used to come to such a conclusion in the book is logically flawed. While it is true that some historical events have been shown to be encoded in the Book of Genesis in certain configurations, it is absolutely not true that every similar configuration of "encoded" words necessarily represents a potential historical event. In fact, quite the opposite is true: most such configurations will be quite random and are expected to occur in any text of sufficient length. Mr. Drosnin states that his "prediction" of the assassination of Prime Minister Rabin is "proof" that the "Bible Code" can be used to predict the future. A single success, regardless of how spectacular, or even several such "successful" predictions proves absolutely nothing unless the predictions are made and evaluated under carefully controlled conditions. Any respectable scientist knows that "anecdotal" evidence never proves anything.[10]

Eliyahu Rips, the coauthor of the original study who had consulted with Drosnin,was also unhappy with the book, though he is featured prominently and favorably in the text. In a public statement, Rips wrote, in part:

> I do not support Mr. Drosnin's work on the codes, nor the conclusions he derives. . . . All attempts to extract messages from Torah codes, or to make predictions based on them, are futile and are of no value. This is not only my own opinion, but the opinion of every scientist who has been involved in original serious codes research.[11]

Other observers have cast doubt on the whole enterprise—not just Drosnin's popularization, but serious code research as well. Australian computer scientist Brendan McKay and collaborators he assembled from Hebrew University, Rips's own institution, have provided a scholarly critique of the original work that disputes its significance.[12] This critique was published by the same journal, *Statistical Science*, that published the Witztum, Rips, and Rosenberg paper. Speaking as he did previously for the editors in an introduction to the McKay article, Robert Kass laments that the publication of the Witztum paper was misinterpreted as a "stamp of scientific approval on the work." He notes that he wrote in his introduction to the paper that it was being offered to readers "as a challenging puzzle." Kass embarrassingly admits that none of the three referees was "willing to spend the time and effort required to reanalyze the data carefully and independently" and assured readers that this has now been done with the McKay article. Kass writes that the new study comprises a "careful dissection and analysis of the Witztum method" and that the "explanations are very convincing and, in broad stroke, familiar." Indeed, they are very familiar to any scientist or mathematician who has done statistical analyses. Kass supports the conclusions of McKay and his collaborators that the protocols used in the original study were insufficiently strict and the statistical analysis flawed. The editor closes by saying that "the puzzle has been solved."[13]

In their article, McKay and his collaborators assert that the Witztum and Rips case is "fatally defective" and the "result merely reflects on the choices made in designing their experiment and collecting data from it." In other words, they saw in the data what they wanted to see. Although Rips and his coauthors insisted that their data sample was strictly preselected, they were not strict enough and allowed sufficient wiggle room to get artifactual positive results.[14]

As you might expect, Rips and Witztum dispute all this.[15] However, the most damaging case against the whole notion of finding messages hidden in the Bible by playing with letters on the computer is that the same can be done with any sufficiently long text, with the same "astounding" results. McKay and his collaborators have provided numerous, amusing examples of how you can find whatever you want, in English as well as Hebrew.[16]

Drosnin had stated, "When my critics find a message about the assassination of a prime minister encrypted in *Moby Dick*, I'll believe them." McKay answered the challenge, using Drosnin's methods on *Moby Dick* to "predict" the assassinations of Indira Ghandi, Martin Luther King Jr., John F. Kennedy, Abraham Lincoln, and Yitzhak Rabin, as well as the death of Princess Diana. Others have joined in the fun. New Mexico physicist Dave Thomas performed an ELS analysis on the King James Version of Genesis and found "Roswell" and UFO" hidden within. He also found "Hitler/Nazi" encoded in Genesis, *War and Peace,* and the 1987 U.S. Supreme Court ruling on *Edwards v. Aguillard,* the creationist decision referred to in chapter 2, which he conveniently had on his computer.[17]

Brendan McKay also reports, on his Web site, that he has found "Brendan McKay," "Chanukah Candles," and "Jesus the Messiah" in *War and Peace.* "Bill Gates" is ominously encoded in the Book of Revelation.

Drosnin claims that the probability of getting many of his results by chance is extremely low. However, it appears he calculated that probability incorrectly, not counting all the opportunities for chance to produce his observed connections in the billions of trials he makes. This mistake is very common. It is called the "file-drawer effect" in statistics and happens to be one that I have often seen made by enthusiastic physicists who are overly anxious to make a major discovery and become famous. Fortunately, the self-correcting procedures in place in physics and other sciences usually catch these human errors before they are announced.

The Bible Code has been widely reviewed, with mixed results. Bible believers have been understandably delighted, and one can find the book promoted by Jewish and Christian organizations. However, I am not aware of any favorable mention of Drosnin's book by mathematics or statistics experts, who are only too conscious of the dangers of being fooled by chance events of low probability when a large number of trials is made. Let me just quote from two reviews by mathematicians that appeared back-to-back in the same issue of *Notices of the American Mathematical Society.*

Allyn Jackson writes that "this book is a series of wild, unfounded claims based on stretching statistical evidence to the breaking point." Drosnin has said he is simply a skeptical journalist in search of the facts, but Jackson notes that he "lacks the mathematical and statistical background that would bring some depth to his skepticism. He is deluded by his own ignorance."[18]

In a following commentary, mathematics professor and Orthodox rabbi Shlomo Sternberg minces no words, writing that "*The Bible Code* by Michael Drosnin exploits a hoax perpetrated by two Israelis, E. Rips and D. Witztum." He points out that the messages found by both the Israeli group and Drosnin would collapse even if just a few letters were added to or dropped from the text they used. As Sternberg explains, "Any serious student of the Talmud knows that there are many citations of the Hebrew Bible which indicate a differing text from the one we have. . . . One of the oldest complete texts of the Bible, the Leningrad codex (from 1009) differs from the Koren version used by Rips and Witztum in forty-one places in Deuteronomy alone. In fact, the spelling in the Hebrew Bible did not become uniformized until the sixteenth century with the advent of a printed version that could provide an identical standard text available at diverse geographical locations."[19]

Drosnin shows his ignorance of biblical scholarship, claiming, in direct contradiction to Rabbi Sternberg, that "every Hebrew Bible that now exists is the same letter for letter."[20] Drosnin also exhibits an ignorance of statistics. He tells us, "The odds against Rabin's full name appearing with the prediction of his assassination were at least 3000 to 1." Perhaps. But then he says, "Mathematicians say 100 to 1 is beyond chance. The most rigorous test ever used is 1000 to 1."[21]

In fact, no mathematician would ever say that "100 to 1 is beyond chance." Recall, from chapter 4, that mathematician William Dembski, in his desire to prove the existence of intelligent design, uses odds of 10^{150} to 1 to define an information sequence as too complex to have happened by chance (although, recall that even this criterion is inadequate after the fact). Furthermore, as we have seen, the threshold used in physics for the publication of a new result is conventionally set at a p-value of 0.01 percent, far more rigorous than the 1000 to 1 or p-value of 0.1 percent that Drosnin tells us is the "most rigorous."

I have gone into the story of *The Bible Code* in considerable detail (and I have still left a lot out) because it provides a good lesson in how not to do science—especially when you are in the business of looking for God by scientific means. As we saw in the preceding section, a claimed prediction is not a meaningful prediction unless it is explicitly published before the fact. Drosnin's "prediction" of Rabin's assassination was probably truthfully made before the fact, although, to my knowledge, it was not published until after the fact. However, we also saw that the only predictions that are worth anything are *risky* predictions. Given the violence of the Middle East, Rabin's violent death was not a very risky prediction. Furthermore, Drosnin admitted that some details were later filled in after the assassination took place.

You only need to read *The Bible Code* to see just how disingenuous Drosnin is on this, his major claim. Throughout the book he talks about how he predicted the correct year of Rabin's assassination, implying this was done prior to his going to Israel to warn the prime minister. Yet Drosnin did not find the year, 5756, on the Hebrew calendar, in association with Rabin's name. Rather, he found it in the vicinity of the words "Holocaust of Israel." This a very tenuous connection, indeed. Any one of a thousand phrases might have given the same association: "Fire in Israel," "Loud noise in Israel," "Sad event in Israel," and so on. Make up your own list. Furthermore, when Drosnin described how this prediction came about, he tells us that he made this discovery "While Israel mourned Rabin," obviously *after* the event.[22]

And this is not all that is fishy about Drosnin's claimed prediction. The expression "Assassin that will assassinate" is his personal translation, and he admits he knew no Hebrew until he started this project. As we have seen, he is hardly a Hebrew scholar. The Jerusalem Bible translates the identical Hebrew text as it appears in Deut. 4:42 as "a man who has killed his fellow."[23] This verse, as translated in the Revised Standard Version, talks about a manslayer killing his neighbor unintentionally. Clearly Drosnin has stretched the facts mightily to say that he predicted the details of the assassination of Yitzhak Rabin in the year it happened, before it happened.

Drosnin makes a number of predictions that, as of this writing, are still in the future, such as an earthquake in Los Angeles in 2010. This would not be a highly unlikely or extraordinary event. A meteor hitting Los Angeles would be a more risky prediction, and I am sure he can find that in the Bible if he looks there or in any other sufficiently long text. As Allyn Jackson notes, Drosnin's prediction of an "atomic holocaust" in Israel in 1995–1996 (the same Hebrew year as Rabin's assassination) did not pan out.[24] However, he later found the word "delayed" encoded nearby. Like most fortune-tellers, Drosnin hedges his bets so that he can win no matter what happens.

As for the low probability of chance claimed for their results by the more credible Rips and his collaborators, McKay and his coauthors demonstrated that this probability was grossly underestimated because account was not taken of the wiggle room they allowed in the rules used to conduct the experiment. Recall that the statistical significance of 1.6×10^{-5} odds quoted in the Rips paper was based on a choice of thirty-two "famous rabbis" whose names and other biographical information were found embedded in the Bible. McKay shows that by removing four not-so-famous ones, the results become statistically insignificant. We should be highly suspicious of any effect that is highly sensitive to the particular sample used. A good experiment must always contain carefully constructed safeguards to prevent the experimenters from, consciously or unconsciously, selecting a sample that gives their desired results and ignoring those samples that do not—the file-drawer effect mentioned earlier. If several different samples are used in the study, then the calculation of chance odds to be presented in the final report must include those that fail to give the effects being claimed as well as those that do.

This is the same problem that arises in the interpretation of an anecdotal claim for some paranormal event, such as a prophetic dream. To calculate the odds that the event was a chance occurrence, we need to know all the times the same dream occurred without coming true. Few people even remember having such dreams.

I have presented only the first round of arguments on both sides of the Bible code debate. The interested reader can follow up on the responses that the various antagonists have made to the other side's rhetoric. However, I doubt that the Bible side will win this one—unless they discover some new, risky prediction, publish it in a respectable scientific journal before the fact, and then have it come true. Based on the situation, as it now exists, I think we can safely draw the same conclusion we did in the preceding section, that no case can be made for the Bible as a source of revelatory information. In November 2002 a sequel, *Bible Code II: The Countdown*, by Michael Drosnin was released by Viking Press predicting a range of catastrophes leading to Armageddon. Dave Thomas has taken the 6,966 characters from book excerpts on the Amazon.com Web site and extracted this secret message: "The Bible Code is silly, dumb, fake, false, evil, nasty, dismal fraud, and snake-oil hoax."

The whole Bible code saga provides yet another example of, and further insight into, the thinking processes that we saw repeatedly in our previous review of the design arguments for the existence of God. Hugh Ross, Michael Behe, William Dembski, and Bible decoders perceive God or "intelligent design" in their data because that's what they want to see. They are like the people who imagine a man in the moon or the face of Jesus on a burned tortilla.

Note that, as was the case with the design arguments discussed earlier, we again find a scientific dispute over how to make proper probability calculations. Ross insists that the fine-tuning of the constants of physics is too improbable to have happened by chance or other natural process. But he does not demonstrate this quantitatively. He simply assumes his readers will find his assertions plausible, since most want to believe them anyway. Behe and Dembski make a like argument, that the order we see in biology is too improbable to be natural. But, again, they can only call upon their readers to take their word for it. The statistical analysis of the Bible code shows that what appears on the surface to be highly improbable can in fact be very likely when you select for the desired outcome. Ironically, that's exactly how natural selection works, with the desired outcome being gene survival. Furthermore, even if Ross, Behe, Dembski, and the Bible decoders are correct in their guesses that the probabilities for certain happenings being natural or purposeless are low, they must also demonstrate why the alternative, nonnatural or purposeful probabilities are not even lower.

TESTING FOR GOD

Let us now move to a quite different arena in which direct, empirical evidence for the reality of the Judeo-Christian-Islamic God has been recently claimed.[25] Indeed, as I have emphasized throughout, it is the existence of this God—one who plays a highly active role in everyday human affairs—and not all conceivable gods, that can be tested scientifically. We simply have to look for the results of all the activity of God that may be reasonably anticipated and attempt to analyze these results scientifically. This should be, in principle, no more difficult than looking for effects of environmental pollution on the growth of corn or sending a beam of neutrinos into a block of matter and studying what comes out of that block.

Whatever other properties the God of Jews, Christians, and Muslims may have, he is a God who answers prayers. As Jesus promised in John 16:23–24: "And in that day ye shall ask me nothing. Verily, verily, I say unto you, Whatsoever ye shall ask the Father in my name, he will give it you. Hitherto have ye asked nothing in my name: ask, and ye shall receive, that your joy may be full." With all the billions of prayers being said every day throughout the world, their effect should be readily measurable.

The task of any empirical research program is to make observations under carefully controlled situations and to try to measure any causal connections between the data and existing or hypothesized theories. Usually, several different explanations for the observations are possible; otherwise, why do the research? Thus, the experiment must be designed to distinguish among the various possibilities. Even when the data seem to support a certain pet theory favored by the experimenters, they cannot immediately conclude that their theory is thereby validated. Other explanations must be ruled out with high probability. And although great care may be taken to eliminate alternative causes, the possibility remains that the result was simply due to chance—a random statistical fluctuation.

Fortunately, the methods of statistical error analysis are now highly developed, and most scientists are trained in their application. Unfortunately, many authors who claim to make scientific statements are not very knowledgeable in these techniques. We saw this in the above discussion of the Bible code. Recall Drosnin's claim that his prediction of the Rabin assassination had an estimated 3000-to-1 odds against being an artifact of chance. This sounded very impressive, and Drosnin asserted that anything above 100 to 1 is "beyond chance," while claiming that 1000 to 1 is the "most rigorous test ever used."

But, as I pointed out, beside Drosnin's misinterpretation of the p-value, these two statements are wrong. Odds of 100 to 1 are hardly beyond chance, and tests more rigorous than 1000 to 1 are commonplace in many fields.

Consider the following simple example, which I have used before. Suppose you have five coins. It would seem very unlikely that they will all fall heads (or tails) up in any given toss. However, they will do so, on average, every 32 tosses. Let us make it more interesting and toss twelve coins. These will fall heads up on average once every 4,096 times. These odds are even greater than Drosnin's 3000 to 1. Now, suppose instead of looking just for some preselected pattern, like all heads up, you simply look for anything interesting. That could include all tails up, alternating heads and tails, the first six heads and the second six tails, and so on. Obviously the chance odds can then decrease to the point where "something interesting" would be commonplace.

This was the situation we saw in the discussion of the intelligent design arguments in chapter 4. In claiming evidence for design in nature, William Dembski proposes that we look for complex specified information. He defines complex information as a minimum of 500 bits. The chance odds for any given string of 500 bits is one in 10^{150}, if the string is specified in advance. This does not mean that 500 bits of information is impossible to generate. You can easily produce 500 bits of previously *unspecified* information. Just write down any sequence of 150 digits from 0 to 9; it will take you only a few minutes. Or, you can write a string of 500 binary digits: 0s and 1s. The probability for the sequence generated, after the fact, is 100 percent!

So it is not just the number of bits, the complexity of the information, but the specification of a particular bit sequence that makes the information highly unlikely to be the results of chance. After you write down your 150 digits, you would not reproduce that sequence again, with a random number generator, in the age of the universe. However, as we saw, Dembski does not limit his definition of specified information to preselected sequences, allowing the postselection of "interesting" patterns. While he claims to provide a prescription for distinguishing suitable postselected patterns (which he calls "specifications") from unsuitable ones (which he calls "fabrications"), it is not clear that any such distinction can be made.

Drosnin's 3000 to 1 seems very unlikely by chance until you remember that he was looking not just for "Yitzhak Rabin" but for "John F. Kennedy," "Robert F. Kennedy," "Anwar Sadat," "Martin Luther King," and who knows how many other familiar names. He was not doing a single experiment, like tossing twelve coins once. He was, in effect, tossing twelve coins thousands of times in his computer and selecting whatever trials came out in what he regarded, after the fact, as an interesting sequence. And this is exactly how evolution produces the complex sequences of information that constitute living organisms: interesting patterns are chosen from random ones by natural selection.

With this lengthy introduction, we are now in a position to evaluate, in a scientific fashion, whether events that are claimed to be affected by supernatural intervention are sufficiently significant statistically so as to make highly unlikely their occurrence being merely the result of chance.

If you were to rely on the personal testimony of millions, you would conclude that the supernatural does indeed exist. Most believers can cite instances in which they or someone they know have prayed for a particular outcome and had it happen, or had other personal experiences that convince them of the existence of God. They will readily admit that many of their prayers have gone unanswered, but this does not, in most cases, weaken their faith. Surely it would be unreasonable for God to answer every prayer, or even most prayers. He answers the ones he wants to answer. These may not be many in number, but not zero if he is the God of Judaism, Christianity, and Islam. Just a few clear cases of prayers being answered would be sufficient to prove the existence of that God, if no other explanation can be found.

However, we cannot make a definitive statement based on anecdotes or other uncontrolled observations. These have too many variables that can affect the outcome. We have no idea how to take these variables into account in our calculation of chance odds. For example, how unlikely is it to have a specific dream come true? We can calculate that only by knowing all the times that specific dream did not come true, and records of unfulfilled dreams are not kept. Normally, they are not even remembered. It is not possible to say that some happening was highly unlikely to be the result of chance without counting all the outcomes. For example, suppose that the odds against a particular individual winning a lottery are 10 million to one. If a million people enter the lottery, then the odds of someone winning are one in ten. With billions of people on the planet, unlikely events will happen every day by chance. That's why we must limit ourselves to controlled experiments or other types of observations in which all the variables are known and we can properly estimate the odds.

Throughout history, and widespread even today, certain select individuals have been reported as having the supernatural power to work miracles. Suffice it to say, none have done so under controlled conditions acceptable to the scientific consensus. Of special note are faith healers who claim to cure in the name of God. Despite thousands of anecdotes, no faith healer has proved his or her ability in scientific trials. Indeed, many have been uncovered as charlatans who prey upon the sick, taking their money while giving them false hopes and sometimes discouraging them from getting proper treatment. Magician James Randi has skillfully exposed many members of this unsavory group, although they still thrive on religious cable channels and at revival meetings often attended by thousands.[26] I cannot understand why religious leaders do not denounce the practice of faith healing and why laws are not passed against it. Surely these shameless individuals do the image of religion no good.

Faith healing, if it were successful, would provide a good case for God. Its lack of success is a good case against. Not only are people robbed of their money and their dignity by unscrupulous faith healers, those who practice faith healing in an honest and pious belief that they are helping others can end up doing more harm than good. For example, pediatricians Seth Asser and Rita Swan have looked at the cases of 172 children who died between 1975 and 1995 whose parents withheld medical care because of reliance on religious faith-healing rituals. They found that 140 fatalities were from conditions for which survival rates with medical care would have exceeded 90 percent. Eighteen more had expected survival rates of greater than 50 percent. All but 3 of the remainder would likely have had some benefit from clinical help. The authors concluded: "When faith healing is used to the exclusion of medical treatment, the number of preventable child fatalities and the associated suffering are substantial and warrant public concern." The authors call for legislative action to protect these innocent victims of fanatical belief.[27]

Hundreds of published scientific reports can be found that claim measurable effects of religious or spiritual actions. Virtually all are found in the health domain, in which the possible therapeutic effects of prayer and other religious activities are investigated.[28]

In general, medical research falls into two categories: (1) *epidemiological*, in which health records of a large population are examined for associations between various characteristics and outcomes; and (2) *experimental*, in which a specific intervention is applied to a sample of patients and compared with a control sample of patients picked from the same set but for which the intervention is not applied. In (2), the patients assigned to each sample are selected randomly, and if the experiment is done properly, neither the patients nor the attending medical personnel know who has been assigned where. That is, the experiment is "blind." Let us take a look at some of the reported results in each of these categories.

THE EPIDEMIOLOGY OF RELIGION

Epidemiological studies, in general, are notoriously difficult to interpret reliably because of so-called confounding factors. A given study may indicate a correlation between an illness and some factor, but this does not necessarily prove that the factor is the cause (or cure) of the illness. In an amusing example given by psychiatrist Richard Sloan, a study might find that lung cancer is more prevalent among people who carry around matches in their pockets. It would not follow that matches cause lung cancer.[29]

Perhaps the first scientist to do an epidemiological study on the possible health benefits of religion was Francis Galton, the prominent nineteenth-century anthropologist (and cousin of Darwin). Galton looked at published data on the mean lifetimes of men in upper classes in England, including members of royal houses, clergy, lawyers, medical doctors, military men, and others. He noted that royalty had the shortest mean lifetime, 64.04 years, despite being the most prayed for. As Galton (1872) interprets this result:

> The sovereigns are literally the shortest lived of all who have the advantage of affluence. The prayer has therefore no efficacy, unless the very questionable hypothesis be raised, that the conditions of royal life may naturally be yet more fatal, and that their influence is partly, though incompletely, neutralized by the effects of public prayers. [30]

Clergy did better, living on average 69.49 years, but that was reduced to 66.51 years when the more prominent members of the class were singled out. Galton remarks: "Hence the prayers of the clergy for protection against the perils and dangers of the night, for protection during the day, and for recovery from sickness, appear to be futile in result." Galton's sample is too small to draw any firm conclusion except to say that he found no evidence for a strong effect of prayer on the lives of people who would be the most expected to benefit.[31]

Recently we have seen an upsurge in epidemiological studies that attempt to discern a connection between religion and health. In a 1998 review published in *Archives of Family Medicine*, Dale Matthews and five collaborators examined a large number of papers and concluded that "a large proportion of published empirical data suggest that religious commitment may play a beneficial role in preventing mental and physical illness, improving how people cope with mental and physical illness, and facilitating recovery from illness."[32]

One of the coauthors of this study, which was partially funded by the Templeton Foundation, was Harold G. Koenig, director of the Center for the Study of Religion/Spirituality and Health located at Duke University. Koenig is also the author of a popular book, *The Healing Power of Faith*, and has appeared on national television promoting the power of religion to heal. In his book he claims to have evidence that people with strong faith are healthier, experience less depression, and live longer.[33]

A review of the literature from a more critical perspective has been provided by R. Sloan, E. Bagiella, and T. Powell in the British medical journal *Lancet*.[34] They concluded that the published work lacks consistency and is not based on sufficiently large samples of data. Linda Gundersen has also reviewed the subject in *Annals of Internal Medicine* and finds that the conclusions of many of the studies are doubtful.[35]

By far the greatest number of these investigations is of the epidemiological variety, in which various religious behaviors are examined for their possible association with health. Churchgoing seems to produce the greatest correlation, but none of these reports have the statistical significance that merit their being regarded as definitive. Furthermore, the studies have been unable to adequately establish a clean causal connection between religion and health. As Sloan and his coauthors explain, when confounding variables are considered, these can explain most if not all of the effects observed. For example, a 1972 study that is often cited as evidence for a positive association between church attendance and health[36] was later found, by one of its own authors, to be mistaken due to failure to control for people with reduced mobility. People in poorer health were simply less likely to go to church.[37] Incidentally, this is a nice example of science successfully policing itself.

A similar question regarding the effects of confounding variables can be raised with the frequently heard claims of a connection between positive thinking and health. Perhaps such a connection exists, but is it positive thinking that causes good health or good health that causes positive thinking?

A much ballyhooed recent report has also been misinterpreted by those who chose not to read carefully what the investigators actually wrote. The autobiographies of 180 catholic nuns were studied, and it was found that those who reported positive emotions in childhood lived longer than those who did not, with a p-value of 0.1 percent.[38] This did not mean that nuns live longer than the rest of us because of supernatural intervention, just that positive-thinking nuns may live longer than negative-thinking nuns.

The best way to extract a single factor for a difference in observations among two samples is for the samples to be equal in every way except that one factor. Thus, for experimental purposes, groups of religious people need to be compared to groups of nonreligious people, with the two groups suitably identical in every other conceivable respect. This should be possible. Not all atheists indulge in behaviors that are damaging to their health.

It might even turn out that the atheists do better! Researchers must be careful to to guard against any bias against that possibility. After all, far fewer prison inmates are atheists in proportion to their representation in the general population, and far more are Christians. But, I must admit, this has a confounding factor, too: parole boards tend to look favorably on convicts who profess to be "born again."

In fact, studies exist that indicate a negative influence of religion on health, although these are not always publicized with the same fanfare as those that give positive effects. As Kevin Courcey has discovered, Koenig and his collaborators selected favorable studies in their review, mentioned above, and ignored even some of their own research that gave conclusions opposite to the ones they preferred to see.[39] For example, a 1994 Koenig paper notes that "several studies have reported an association between psychiatric disorder and religious affiliation, with the rates of disorder highest among nonmainline Protestant religious groups."[40] Another Koenig study ignored in his review showed that the likelihood of major depression among highly religious Pentecostals was three times greater than among persons who said they had no religious preference.[41] Other investigators have also shown that the religious are frequently among the lowest in mental health.[42] Koenig also fails to mention other studies on the role of religion in child abuse and medical neglect.[43]

Another recent report indicates that religion may not be so good for you after all. As reported in the *Archives of Internal Medicine,* Bowling Green State University psychologist Kenneth Pargament and his colleagues, which included Koenig, looked at 595 individuals aged fifty-five or older who had been hospitalized between 1996 and 1997. Over 95 percent were conservative or mainline Christians. Jews, Buddhists, or nonbelievers were not studied. Patients who reported feeling alienated from God or believing the devil brought about their illness were associated with up to a 28 percent increase in risk of dying during the two-year follow-up period. None of the other factors considered, such as gender or race, showed any significant difference.[44] Of course, this could simply be the effect of negative thinking. I mention this only to show that the effect of religion on health is ambiguous at best and that the one-sided reports one usually reads in the media provide no scientific support for a supernatural role in heath.

Even if some positive correlation between religion and health can be found in some areas (mental health does not seem to be one of them), religious people, as a group, are less likely than the general population to engage in risky behaviors such as smoking, excessive alcohol consumption, and promiscuous sex. One can conceive of some value of religious behavior without invoking divine interventions as necessary causes.

For years, Herbert Benson has been studying the health benefits of prayer, meditation, and relaxation.[45] The statistical significance of some of his claimed positive benefits have been questioned.[46] But even if the reported effects are real, are they spiritual? Do they provide evidence for a spirit realm? Hardly. The ability of the mind to heal or hurt the body is well known. It seems perfectly reasonable and natural that relaxation by any means can have a benefit, for example, by lowering blood pressure and reducing stress. Benson provides no evidence that a Christian prayer works better than a Hindu one or a simple, perfectly secular relaxation exercise.

INTERCESSORY PRAYER

Praying for yourself might help you. Or it might harm you. Either outcome is possible, by purely material brain-body interactions with nothing spiritual required. And praying for another person with his or her knowledge might also help, by purely material means, in reducing that person's stress. Or it might hurt by applying more stress. Believe it or not, prayer may be harmful! Scott Walker and his collaborators have reported on the results of remote prayer on the patients of an alcohol treatment facility. Their conclusion: "Compared with a normative group of patients treated at the same facility, participants in the prayer study experienced a delay in drinking reduction. Those who reported . . . that a family member or friend was already praying for them were found to be drinking significantly more at six months than were those who reported being unaware of anyone praying for them."[47] However, based on the data, we have to conclude that prayer does little or nothing most of the time.

We can safely put aside all the claims for evidence of a God who answers prayers in the set of epidemiological studies published to-date. Nevertheless, no material explanation is likely to be able to account for intercessory prayer done blindly at a distance—should it ever be successfully demonstrated. In another widely reported investigation (the negative reports, like those mentioned above, rarely make the media), cardiologist Randolph Byrd published an article in a regional medical journal claiming evidence that coronary patients benefitted from blind, distant intercessory prayer.[48] His patient group was comprised of 192 who were prayed for by born-again Christians from a local church. The control group contained 201 patients. It was claimed that this was a double-blind experiment, with subjects randomly assigned to treatment group or control group, and with treatment staff uninformed of the assignments.

As treatments progressed, the patients in both groups were graded in twenty-six medical categories as "good," "intermediate," or "bad." In six of these, the prayed-for group did better at the level of 5 to 7 percent. In one of the six categories, for example, 7 percent of the patients needed fewer antibiotics. But no significant difference was found in total days in the hospital or death rates, despite intercessory prayer specifically requesting a "rapid recovery."

How significant is the reported improvement? Byrd himself admitted in the paper: "Even though for [the six seemingly significant] variables the P values were less than 0.05, they could not be considered statistically significant because of the large number of variables examined. I used two methods to overcome this statistical limitation . . . [the] severity score, and multivariant [sic] analysis."[49]

In other words, the results were insignificant. However, like so many others working in this field, Byrd said something quite different to the press. The *Kansas City Star* reported him as stating unequivocally that "the patients who were prayed for just did better."[50] Has anyone ever told you not to believe everything you read in the papers (or hear on radio, see on television, or read on the Internet)?

Several further comments are called for at this point. As I have noted, the criterion of p-value less than 0.05 that is often accepted in medical journals is far from adequate for such an extraordinary claim. Byrd, in order to get published in a scientific journal, was forced to acknowledge that his result is not significant when all variables are considered. He claims to improve this with a "multi-variant" analysis (the correct term is "multivariate"). However, decades of experience in experimental physics have taught me that no complex statistical analysis will give convincing results when those results are not already significant based on a simpler, straightforward analysis. The only honest way to improve poor statistics is with more data.

A detailed critique of the Byrd experiment has been given by internist Gary Posner. He writes, "The most striking flaw in this study's methodology is one also forthrightly acknowledged by Byrd. 'It was assumed that some of the patients in both groups would be prayed for by the people not associated with the study; this was not controlled for.'"[51]

I am not as concerned as Posner about this point. It is true that apologists can use this to "explain" why the experiment did not give strong evidence for the efficacy of prayer, and this makes the prayer hypothesis nonfalsifiable and thus not very scientifically credible. However, suppose that the prayed-for group had improved so significantly that the chance explanation had been very unlikely, say with a p-value of 0.01 percent. Then we would have good evidence that the prayers of born-again Christians are not only effective but far more effective than your run-of-the-mill prayer, say by a Methodist or a Catholic. The experiment did not produce this profound result, of course, but its hypothetical possibility at least demonstrates that a scientific study of the efficacy of prayer can be, in principle, definitive.

A more serious flaw in the methodology of the Byrd experiment has been uncovered by Irwin and Jack Tessman. They discovered that the experiment was not blind after all. The code used to keep the experimental group indistinguishable from the control group was broken when Byrd wrote the first manuscript of his paper. In order to answer criticisms and produce a new version of his manuscript, he had to decide again which group did better, and this was now done unblinded. Furthermore, it was revealed that the coordinator of the study, who kept the detailed records on the patients, was completely unblinded all along.[52]

The media hype surrounding Byrd's results undoubtedly led millions to think that science had demonstrated that prayer works, and, by implication, that God exists. However, these conclusions can hardly be drawn from this abysmal piece of work.

In 1999 another study along the same line as Byrd's was published in a major medical journal, *Archives of Internal Medicine*, by W. S. Harris and nine coauthors.[53] They report that the length of stay in the coronary care unit decreased by 9 percent in the prayer group, but with very low statistical significance. The p-value for their result, according to their own calculation, was over 0.25. The total length of stay in the hospital actually increased by 9 percent for the prayer group, but this was also insignificant, with a p-value of 0.25.

Using a standard method for scoring cardiac treatment outcomes based on thirty-four adverse conditions, the investigators obtained an 11 percent advantage for the prayer group with a p-value of 0.04. Although the paper was published, the authors admitted, however, that their result could be simply chance.

When Harris and his collaborators applied the Byrd criteria, they obtained insignificant differences, with a p-value of 0.33 compared to Byrd's reported 0.05. If one was to take these statistical estimates seriously, you would have to conclude that Harris and his collaborators have disconfirmed Byrd. A more reasonable conclusion is that neither experiment showed anything significant, one way or the other.

In an amusing postscript to these studies, psychologist Nicholas Humphrey has looked at the Harris results and discovered a far more significant effect in the data than that reported by the authors. The original sample contained 1,013 patients, and these were all randomized to determine who would and who would not be prayed for. However, the praying took a day to organize, so the researchers removed from their sample those patients who were released in less than twenty-four hours, leaving 990 in the two care groups. Humphrey found that 18 of the 484 patients who were going to be prayed for were released—before the praying started—while only 5 of the 529 who where not going to be prayed for left the hospital in that first day. The p-value for this difference, according to Humphrey, is 0.1 percent! He concludes that "either the study has come up with strong evidence of prayer producing *backward causation* of recovery, or else there was something wrong somewhere with the way the study was conducted."[54]

Actually, they were never prayed for, so it can't be attributed to backward causation, which Humphrey obviously proposed tongue-in-cheek. I conclude that we have yet another example of how statistics can fool you when you analyze data in all sorts of combinations until you find something that looks interesting. Once again we see the filedrawer effect in action.

After the above was written and submitted to the publisher, another study on the effect of intercessory prayer on coronary patients was reported. The abstract reads:

> The results of twenty-six weeks of intercessory prayer, a widely practiced complementary therapy, were studied in 799 patients randomized to an intercessory prayer group or to a control group after discharge from a coronary care unit. As delivered in this study, intercessory prayer had no significant effect on specifically defined medical outcomes, regardless of risk status.[55]

THE TRAGIC STORY OF ELISABETH TARG

Clearly we must be highly skeptical of any scientific claim that is made by "true believers," whether the beliefs being tested are religious or secular. Even the most sincere investigators may unconsciously select data that support their beliefs and ignore the data that do not, when those beliefs are deeply held. Yet another, very tragic, story needs to be mentioned to emphasize this point.

In 1998 psychiatrist Elisabeth Targ and her collaborators published a paper in the *Western Journal of Medicine* which claimed that various forms of "distant healing," including prayer and "psychic healing," significantly improved the health of patients with advanced AIDS. They reported on what they claimed was a "double blind, randomized trial" involving forty patients in the San Francisco area. Here is how the authors summarized their results, with *p*-values (*P* =) indicated:

> At six months, blind medical chart review found treatment subjects acquired significantly fewer new AIDS-defining illnesses (0.1 vs. 0.6 per patient, *P* = 0.04), lower illness severity (severity score 0.8 vs. 2.65, *P* = 0.03), required significantly fewer doctor visits (9.2 vs. 13.0, *P* = 0.01), fewer hospitalizations (0.15 vs. 0.6, *P* = 0.04), and fewer days of hospitalization (0.5 vs. 3.4, *P* = 0.04). Treated subjects also showed significantly improved mood compared to controls (change in POMS [a measure of mood] –26 vs. +14, *P* = 0.02). There were no significant differences in CD4+ counts.[56]

The fascinating tale surrounding this experiment and its aftermath was related in a recent article by Po Bronson in *Wired* magazine.[57] Targ was the daughter of famous parapsychologist Russell Targ who conducted experiments on extrasensory perception in the 1970s that were published in the journal *Nature*.[58] Russell is a true believer and Elisabeth grew up with a firm conviction that the mind possesses paranormal powers.[59] Indeed, Russell made much of her apparent psychic powers in his 1985 book *The Mind Race*, written with Keith Harary.[60] While obtaining a degree in conventional psychiatry, the paranormal remained of interest to Elisabeth and she found willing support and collaboration from parapsychologists to conduct studies in distant healing.

As a direct result of her promising 1998 study, published in a peer-reviewed journal, Targ received $1.5 million in grants from the U.S. National Institutes of Health's Center for Complementary and Alternative Medicine. The grants provided support for continuing distant healing studies on AIDS and further work on the brain cancer Glioblastoma multiforme (GBM). While rare, GBM is one of the most malignant forms of cancer with survival rates of about a year, even with the most drastic medical intervention. Patients were dying, and it seemed that anything was worth a try.

By an extraordinary coincidence, in early 2002 Targ was diagnosed to suffer from GBM. Surgery was unable to excise all the cancer and, understanding full well her slim chances of recovery by any medical means, Targ was unsure she wanted chemotherapy.

As word of Targ's illness spread, healing groups worldwide began to pray for her. As Bronson describes it:

> Her bedroom turned into a circus. Healers from everywhere showed up wanting to help. It was rarely peaceful and quiet. There was Phillip Scott, a Lakota sun dancer who burned sage; Nicolai Levashov, a Russian psychic who waved his hands; Harriet Bienfield, an acupuncturist with rare Chinese herbs; Desda Zuckerman, an energy worker who used techniques inspired by the ancient methods of the Miwok peoples. The Reverend Rosalyn Bruyere phoned often, trying to get on Targ's schedule. And, of course, there was her father, Russell, urging her to meditate, calm her mind, and go to that place.[61]

Sadly, neither science nor spirit was able to save Elisabeth Targ.

This tragic tale does not end with Targ's untimely death at the age of forty. Bronson reports that the AIDS study published in the *Western Journal of Medicine* had not been blinded—a fact not known to the journal. He says he has confirmed the following from one of Targ's coauthors, biostatistician Dan Moore, and physicist Mark Coings, who married Targ shortly before her death. Although not mentioned in the publication, the original study was designed to look at mortality rates. When Moore broke the code, he found that only one subject had died and so the mortality data were meaningless. Targ and the other collaborators insisted that Moore examine the data further and look at certain HIV physical symptoms and quality of life. He found that the treatment group did no better than the control group. In fact, in some cases they seemed to do worse. Targ urged him to keep looking. Finally, after more data mining, Moore found that the treatment group had "statistically significant" fewer hospital stays and doctor visits. Targ was at a conference at the time and excitedly announced the results.

However, as Bronson put it:

> This isn't what science means by double-blind. The data may all be legitimate, but it's not good form. Statisticians call this the sharpshooter's fallacy—spraying bullets randomly, then drawing a target circle around a cluster. When Targ and Sicher wrote the paper that made her famous, they let the reader assume that all along their study had been designed to measure the twenty-three AIDS-related illnesses—even though they're careful never to say so. They never mention that this was the last in a long list of endpoints they looked at, or that it was data collected after an unblinding.[62]

If Bronson's report is correct, the authors misrepresented their experiment, claiming they had done a blinded study and implying that their criteria were preselected, when, in fact, they searched many postselected criteria until they found what they wanted to see.

The editors and referees of the journals involved in the studies I have reported have done a great disservice to both science and society by allowing these highly flawed papers to be published. This has given a false scientific credibility to the assertion that prayer or other spiritual techniques work miracles, and several best-selling books have appeared that exploit that theme. Telling people what they want to hear, these authors have made millions.

HEALING WORDS

There is yet another angle to the story of prayer and healing. In his popular book *Healing Words*, physician Larry Dossey tells about the search he conducted in which he found "an enormous body of evidence: over one hundred experiments exhibiting the criteria of 'good science,' many conducted under stringent laboratory condi-

tions, over half of which showed that prayer brings about significant changes in a variety of living beings."[63] One wonders why he
would even count those that were not conducted under stringent
laboratory conditions. In any case, "many" purportedly were.

Dossey uses as his main data source a survey by Daniel J.
Benor, M.D., published in the journal *Complementary Healing
Research*.[64] Dossey describes and summarizes Benor's results as
follows:

> Benor defined healing as the *intentional influence of one or more
> people upon another living system without utilizing known physical
> means of intervention.* His findings:
>
> - Researchers have performed 131 controlled trials.
> - Fifty-six of these show statistically significant results at a
>   probability level of < 0.01 or better (that is, the likelihood
>   that the results were due to chance was less than 1 in 100).
> - Twenty-one studies demonstrate results at a probability
>   level of 0.02 to 0.05 or better (that is, the likelihood that the
>   results were due to chance was between 2 and 5 chances in
>   100).
> - These experiments deal with healing effects on enzymes,
>   cells, yeasts, bacteria, plants, animals, and human
>   beings.[65]

Dossey informs us that "ten of the studies are unpublished doc-
toral dissertations, two are masters' theses, and the rest are published primarily in parapsychological journals." He claims that
"these publications have peer review standards as rigorous as
many medical journals." Note that Dossey, as quoted above, also
misinterprets the meaning of *p*-values.

I have already made clear what I think of the standards of medical journals, allowing that these can be somewhat low so as to
make sure that useful therapies are not kept from needy patients
for too long. Unlike physicians, parapsychologists are not in the
business of saving lives but rather that of investigating extraordinary phenomena. In *Physics and Psychics* I argued that those who
search for psi phenomena should be bound by the stricter stan-

dards of physics, which also deal with extraordinary phenomena. If these standards were applied, none of the 131 trials mentioned above would be publishable.

Dossey is certainly wrong when he interprets the above as "simply overwhelming [evidence] that prayer functions at a distance to change physical processes in a variety of organisms, from bacteria to humans."[66] Even without examining the detailed protocols of these experiments, the statistical significance is simply insufficient to draw such a conclusion. We have no idea how many experiments may have been done that gave no positive effects and consequently were never published. As with the Bible code, do enough trials and you will eventually get to see what you want to see.

Dossey acknowledges the problems of dealing with human subjects, which can make the interpretation of experimental data very difficult. How can one get a pure sample of people who no one prays for? The pope prays for everybody, even atheists. My Christian friends pray for me, both for my health and in the hope that I will eventually "see the light." Dossey thinks that the answer is to focus on nonhuman subjects: bacteria, mice, and even cells. However, even in this case, as Hector Avalos points out, there are people praying for the well-being of all life on Earth.[67]

As we have seen, because of the difficulty of obtaining a pure, prayer-free sample, the prayer hypothesis does not have the feature of falsifiability that lends considerable credibility to any scientific hypothesis. However, not all scientific hypotheses are falsifiable, so it would be unfair for me to rule out the prayer hypothesis on that basis. What this means, in practice, is that the type of negative results we have seen so far cannot be used to *prove* that prayer does not work. On the other hand, convincing positive results with humans or other life-forms could have, in principle, provided a good indication for some kind of nonnatural force in action. This has not happened.

NOTES

1. Peter W. Stoner, *Science Speaks: An Evaluation of Certain Christian Evidences* (Wheaton, Ill.: Van Kampen Press, 1952); Josh McDowell, *Evi-*

dence That Demands a Verdict (San Bernardino, Calif.: Here's Life Publishers, 1972); Douglas Geivett and Gary Habermas, eds., *In Defense of Miracles: A Comprehensive Case for God's Action in History* (Downers Grove, Ill.: InterVarsity Press, 1997).

2. The fact that the neutrinos from the Sun have a lower intensity than predicted by the standard solar model is a long-standing puzzle. It now appears that this results from a phenomenon called "neutrino oscillation," a subject I personally worked on for many years and played a small role in its eventual demonstration. This serves as a good example of how the failure of a prediction in physics helps us gain new knowledge as we search for explanations of the failure.

3. For my more detailed discussion of this example, see Victor J. Stenger, *Physics and Psychics: The Search for a World Beyond the Senses* (Amherst, N.Y.: Prometheus Books, 1990).

4. "False Prophecies, Broken Promises, and Misquotes" [online], www.skepticsannotatedbible.com/prophecy.html. Thanks to Bob Phillipoff and Farrell Till for pointing out these particularly striking examples.

5. William Harwood, *Mythology's Last Gods* (Amherst, N.Y.: Prometheus Books, 1992), pp. 169–70.

6. Doron Witztum, Eliyahu Rips, and Yoav Rosenberg, "Equidistant Letter Sequences in the Book of Genesis," *Statistical Science* 9 (1994): 429–38.

7. Maya Bar-Hillel, Dror Bar-Natan, and Brendan McKay, "The Torah Codes: Puzzle and Solution," *Chance* 11, no. 2 (1998): 13–19.

8. Robert E. Kass, editorial comment on Witztum et al, *Statistical Science* 9, no. 3 (1994): 306.

9. Michael Drosnin, *The Bible Code* (New York: Simon and Schuster, 1997).

10. Gans, Harold, "Public Statement" [online], members.nbci.com/bcodes/public2.htm.

11. Rips, Eliyahu, "Public Statement" [online], members.nbci.com/_XOOM/bcodes/public.htm.

12. Brendan McKay, Dror Bar-Natan, Maya Bar-Hillel, and Gil Kalai, "Solving the Bible Code Puzzle," *Statistical Science* 14, no. 2 (1999): 150–73.

13. Robert E.Kass, "Introduction to 'Solving the Bible Code Puzzle' by Brendan McKay, Dror Bar-Natan, Maya Bar-Hillel, and Gil Kalai," *Statistical Science* 14, no. 2 (1999): 149.

14. McKay et al., "Solving the Bible Code Puzzle."

15. A Web search here will lead the reader to many sites promoting all sides of this issue (there are more than two) and provide considerable insight into personal motivations.

16. Brendan McKay, Web site [online], cs.anu. edu.au/people/bdm.

17. David E. Thomas, "Hidden Messages and the Bible Code," *Skeptical Inquirer* 21, no. 6 (1977): 30–36.

18. Allyn Jackson, review of *The Bible Code*, by Michael Drosnin, *Notices of the American Mathematical Society* 44, no. 8 (1997): 935–38.

19. Shlomo Sternberg, "Comments on *The Bible Code*," by Michael Drosnin, *Notices of the American Mathematical Society* 44, no. 8 (1997): 938–39.

20. Drosnin, *The Bible Code*, p. 38.

21. Ibid., p. 28.

22. Ibid., p. 53.

23. Bar-Hillel et al. "The Torah Codes: Puzzle and Solution."

24. Jackson, review of *The Bible Code*; Drosnin's prediction taken from *The Bible Code*, p. 56.

25. Patrick Glynn, *God: The Evidence* (Rocklin, Calif.: Prima Publishing, 1997).

26. James Randi, *The Faith Healers* (Amherst, N.Y.: Prometheus Books, 1987).

27. Seth M. Asser and Rita Swan, "Child Fatalities from Religion-motivated Medical Neglect," *Pediatrics* 101, no. 4 (1998): 625–29.

28. Dale A. Matthews, Michael E. McCullough, David B. Larson, Harold G. Koenig, James P. Swyers, and Mary Greenwold Milano, "Religious Commitment and Health Status: A Review of the Research and Implications for Family Medicine," *Archives of Family Medicine* 7, no. 2 (1998): 118–24.

29. Richard P. Sloan, "Religion, Spirituality, and Medicine," *Freethought Today* 17, no. 2 (2000): 10–12.

30. Francis Galton, "Statistical Inquiries into the Efficacy of Prayer," *Fortnightly Review* 12 (1872): 125–35; [online], www.abelard.org/galton/galton.htm.

31. Physicist and Anglican priest John Polkinghorne has suggested that royalty lived shorter lives in the nineteenth century because they were more exposed to the medical care of the time.

32. Matthews et al. "Religious Commitment and Health Status."

33. Harold G. Koenig, *The Healing Power of Faith: Science Explores Medicine's Last Great Frontier* (New York: Simon and Schuster, 1999).

34. R. P. Sloan, E. Bagiella, and T. Powell, "Religion, Spirituality, and Medicine," *Lancet* 353, no. 9153 (1999): 664–67.

35. Linda Gundersen, "Faith and Healing," *Annals of Internal Medicine* 132, no. 2 (2000): 169–72.

36. G. W. Comstock and K. B. Partidge, "Church Attendance and Health," *Journal of Chronic Disease* 225 (1972): 665–72.

37. G. W. Comstock and J. A. Tonascia, "Education and Mortality in Washington County, Maryland," *Journal of Health and Social Behavior* 18 (1977): 54–61.

38. Deborah D. Danner, David A. Snowdon, and Wallace V. Friesen, "Positive Emotions in Early Life and Longevity: Findings from the Nun Study," *Journal of Personality and Social Psychology* 80, no. 5 (2001): 804–13.

39. Kevin Courcey, "Religiosity and Health," *Scientific Review of Alternative Medicine* 3, no. 2 (1999): 70–76.

40. Harold G. Koenig, Linda K. George, Keith G. Meador, Dan G. Blazer, and Peter Dyck, "Religious Affiliation and Psychiatric Disorder among Protestant Baby Boomers," *Hospital and Community Psychiatry* 45, no. 6 (1994): 586–96.

41. Harold G. Koenig et al., "Religious Affiliation and Major Depression," *Hospital and Community Psychiatry* 43, no. 12 (1992): 1204.

42. R. W. Hood, "The Usefulness of the Indiscriminately Pro and Anti Categories of Religious Orientation," *Journal for the Scientific Study of Religion* 17 (1978): 419–31; C. E. Ross, "Religion and Psychological Distress," *Journal for the Scientific Study of Religion* 29 (1990): 236–45.

43. B. Bottoms et al., "In the Name of God: A Profile of Religion-Related Child Abuse," *Journal of Social Issues* 51, no. 2 (1995): 85–111; D. Capps, "Religion and Child Abuse: Perfect Together," *Journal for the Scientific Study of Religion* 31 (1992): 1–14; M. A. Persinger, "'I Would Kill in God's Name': Role of Sex, Weekly Church Attendance, Report of a Religious Experience, and Limbic Liability," *Perception and Motor Skills* 85, no. 1 (1997): 128–30.

44. Kenneth I. Pargament, Harold G. Koenig, Nalini Tarakeshwar, and June Hahn, "Religious Struggle as a Predictor of Mortality among Medically Ill Elderly Patients," *Archives of Internal Medicine* 161, no. 15 (2001): 1881–85.

45. Herbert Benson, *Timeless Healing* (New York: Fireside, 1996).

46. Irwin Tessman and Jack Tessman review of *Timeless Healing, Science* 276, no. 2 (1997): 369–70.

47. S. Walker, J. S. Tonigan, W. R. Miller, S. Corner, and L. Kahlich, "Intercessory Prayer in the Treatment of Alcohol Abuse and Dependence. A Pilot Investigation," *Alternative Therapies in Health and Medicine* 3, no. 6 (1997): 79–86.

48. Randolph C. Byrd, "Positive Therapeutic Effects of Intercessory Prayer in a Coronary Care Unit Population," *Southern Medical Journal* 81, no. 7 (1988): 826–29.

49. Ibid.

50. E. Adler, "Prayer Helps Sick in Mysterious Ways, Study Suggests," *Kansas City Star*, October 24, 1999.

51. Gary P. Posner, "God in the CCU?" *Free Inquiry* 10, no. 2 (1990): 44–45.

52. Irwin Tessman and Jack Tessman, "Efficacy of Prayer: A Critical Examination of Claims," *Skeptical Inquirer* 24, no. 2 (2000): 31–33.

53. W. S. Harris, M. Gowda, J. W. Kolb, C. P. Strychacz, J. L. Vacek, P. G. Jones, A. Forker, J. H. O'Keefe, and B. D. McCallister, "A Randomized, Controlled Trial of the Effects of Remote, Intercessory Prayer on Outcomes in Patients Admitted to the Coronary Care Unit," *Archives of Internal Medicine* 159 (1999): 2273–78.

54. Nicolas Humphrey, "The Power of Prayer," *Skeptical Inquirer* 24, no. 3 (2000): 61.

55. Jennifer M. Aviles et al., "Intercessory Prayer and Cardiovascular Disease Progression in a Coronary Care Unit Population: A Randomized Controlled Trial," *Mayo Clinic Proceedings* 76, no. 12 (2001): 1199–1203.

56. Fred Sicher et al., "A Randomized Double-Blind Study of the Effect of Distant Healing in a Population with Advanced AIDS," *Western Journal of Medicine* 169, no. 6 (1998): 356–63.

57. Po Bronson, "A Payer Before Dying," *Wired* 10, no. 12 (December 2002).

58. Victor J. Stenger, *Physics and Psychics: The Search for a World Beyond the Senses* (Amherst, N.Y.: Prometheus Books, 1990), pp. 186–92.

59. Martin Gardner, "Distant Healing and Elisabeth Targ," *Skeptical Inquirer* 25, no. 2 (2001): 12–14.

60. Russell Targ and Keith Harary, *The Mind Race: Understanding and Using Psychic Abilities* (New York: Balantine Books, 1985), pp. 103–16. These page numbers refer to the paperback edition not the original hardcover edition published by Villard Books, 1984.

61. Bronson, "A Prayer Before Dying."

62. Ibid.

63. Larry Dossey, *Healing Words: The Power of Prayer and the Practice of Medicine* (San Francisco: Harper, 1993), p. xv.

64. Daniel J. Benor, "Survey of Spiritual Healing Research," *Complementary Medical Research* 4, no. 1 (1990): 9–33.

65. Dossey, *Healing Words*, p. 211.

66. Ibid., p. 2.

67. Hector Avalos, "Can Science Prove that Prayer Works? The Real Story Behind the Hype," *Free Inquiry* 17, no. 3 (1997): 27–31.

TEN

THE BREATH OF GOD

We stand at the end of the Age of Reason. A new era of the magical expla-nation of the world is rising.

—Adolf Hitler, ca. 1936

A SEPARATE REALM

W e move now to another class of empirical data that is fre-quently put forth as scientific evidence for a transcendent reality beyond the material reality that is revealed by our senses and the instruments of science. This data set has to do with our own inner thoughts and consciousness, and the supposed powers of the mind to subordinate the laws of physics.

Most people believe that their thoughts themselves reveal a spiritual realm that goes beyond matter. The notion of incorporeal spirit, as currently perceived in the West, is largely founded on a religious tradition that is not especially drawn from the Bible, although it has been read into it. Such a notion was probably a common, prehistoric perception held by humans throughout the ages, suggested by their thoughts and dreams. A dead person comes alive again inside one's head. The laws of nature place no limitations on the content of one's fantasies.

By contrast, the wide range of collective observations by thousands of scientists worldwide, using the best instrumentation of modern technology, finds no sign of substances other than matter. This includes the ingredients of mental activity, which seems to be solely the product of material processes in the brain. No cognitive data or theories currently require the introduction of either supernatural forces or immaterial substances such as "spirit."

Supernatural or paranormal beliefs do not merge easily with the new synthesis of physics, cosmology, biology, and neuroscience. These beliefs assert something more—that the universe contains constituents other than those derived from the known particles and forces of physics. By itself, of course, this does not imply a contradiction. Perhaps nonnatural effects and immaterial substances are so tiny as to be very difficult to detect. Or perhaps the detectors of physics and astronomy are not suitable for these phenomena, just as a telescope cannot see bacteria.

I should emphasize that many parapsychologists do not regard psychic phenomena, if they exist, as necessarily supernatural. Perhaps psychic forces are perfectly natural. In that case, their discovery would not automatically imply the existence of the spiritual world of religious tradition. Still, until the nature of psychic forces is uncovered, a spirit world will remain one very popular interpretation of the phenomenon, since it would imply the overthrow of materialism. Nevertheless, only when the evidence for special powers of the mind becomes sufficient for us to take it seriously need we concern ourselves whether they are natural or supernatural. Something that does not exist requires no label.

Observations reported by parapsychologists that purport to demonstrate the psychic powers of the mind are not made with high-tech instruments. Rather, they relate to unusual human experiences, reported as anecdotes, or simple experiments that, in many cases, can be done on the dining table in your own home. Even the experiments conducted in the handful of university parapsychology laboratories are crude and low-tech by the standards of the conventional science being done on the same campuses.

To be found so easily and cheaply, psychic forces, if they exist, must exert powerful control over normal matter. If paranormal claims are valid, then the picture of the universe that has evolved over the last few centuries and continues to be confirmed in the major laboratories and observatories around the world is wrong or, at best, grossly incomplete.

Although not always described so explicitly, mind-over-matter and mind-to-mind communication are imagined as resulting from the flow of "psychic energy," a form of energy not presently registered in the scientific inventory. Perhaps the stars control our lives by means of the transmission of "cosmic energy." Acupuncture, therapeutic touch, and other complementary healing techniques are supposed to work by bringing the body's "vital energy" into better balance. Similarly, well-understood electromagnetic phenomena, such as infrared and Kirlian"auras," are promoted as evidence for a human "energy field." Energy seems to be a unifying concept among many paranormal claims. But energy, as used in physics, is a property of matter, so its meaning in a nonmaterial context is at best metaphorical.

The prevalent belief among peoples East and West is that the human being is composed of both a material body and an immaterial spirit that is often called the *soul*. In European thinking, thanks largely to René Descartes, the soul is also associated with the *mind*. In Asian thinking, which is now surfacing in New Age communities in the West, the soul is imagined as being manifested in a postulated nonphysical human energy field. Much of complementary and alternative medicine, as now widely practiced in the United States as well as in Asia, is grounded on vitalism—the notion that living organisms possess some distinctive quality, an *élan vital*, which gives them that special property we call life. In principle, this could still be material in nature, in which case it would not concern us here where we are considering scientific arguments for a world beyond matter. However, the so-called human energy field is often associated with the spirit or the soul, and if science could find convincing evidence for it, the notion of a spirit world would receive a significant boost.

THE SEARCH FOR LIVING ENERGY

Belief in the existence of a living force is ancient and remains widespread to this day. Called *prana* by the Hindus, *qi* or *chi* by the Chinese, *ki* by the Japanese, and ninety-five other names in ninety-five other cultures,[1] this substance is said to constitute the source of life and mind. In 1939 Richard Wheeler reviewed the history of vitalism in the West and defined it as "all the various doctrines which, from the time of Aristotle, have described things as actuated by some power or principle additional to mechanics and chemistry."[2] Early-twentieth-century theories of vitalism include those of Hans Driesch and Henri Bergson.[3]

In ancient times the vital force was widely identified with breath, which the Hebrews called *ruach*, the Greeks *psyche* or *pneuma* (the breath of the gods), and the Romans *spiritus*. In Genesis, God breathes life into Adam. Modern supernaturalists seem to be saying that they can feel the "breath of God" upon their cheeks. As breath was gradually acknowledged to be a material substance, invisible but physical, like the wind, words such as "psychic" and "spirit" evolved to refer to the assumed nonmaterial and perhaps supernatural medium by which organisms gain the qualities of life and consciousness. The idea that matter alone can do this has never proved popular, and ancient materialist ideas are largely lost in history.

Chi remains the primary concept in traditional Chinese medicine, still widely practiced in China and experiencing an upsurge of interest in the West. *Chi* is energy that is said to flow rhythmically through so-called meridians in the body. The methods of acupuncture and acupressure are used to stimulate the flow at special acupoints along these meridians, although their location has never been consistently specified and differs from practitioner to practitioner. *Chi* energy is not limited to the body, but is believed to flow throughout the environment.[4] When building a house or a major building, many believers rely on a *feng shui* master to decide on an orientation that is well aligned with this flow and to decorate the interior so as to separate the good *chi* from the bad. Of course, practitioners swear that the evidence for the value of *feng shui* is overwhelming, and they carry with them the testimonials of satisfied customers to prove it.

As modern science developed in the West and the nature of matter was gradually uncovered, a few scientists sought evidence for the nature of the living force. After Newton had published his laws of mechanics, optics, and gravity, he spent many years looking for the source of life in alchemic experiments. His search was not irrational, given the knowledge of the day. Newtonian physics provided no known basis for the complexity that is necessary for any purely material theory of life or mind. This would have to wait until the emergence of quantum physics, modern chemistry, and evolutionary biology.

Furthermore, Newtonian gravity had an occult quality about it, with its invisible action at a distance. Gravity seemed to be transmitted instantaneously across space with no intervening matter evident. Perhaps the forces of life and thought had similar immaterial properties. Still, Newton and others who followed the same trail never managed to discover a special substance of spirit.

In the eighteenth century, Anton Mesmer (d. 1815) imagined that magnetism was the universal living force and treated patients for a wide variety of ills with magnets, a therapy still being promoted today (you can find "therapeutic" magnets in most drug stores). The claimed effect of magnetism on red blood cells has been refuted by any number of grade school science fair projects, but you can find many people who still swear by it.[5]

Mesmer believed that "animal magnetism" resided in the human body and could be directed into other bodies. Indeed, patients would exhibit violent reactions when Mesmer directed his energy toward them with the point of a finger, until the flow of "nervous current" would rebalance the patient's energies.[6]

In the late nineteenth century, prominent scientists such as physicists William Crookes and Oliver Lodge sought scientific evidence for what Crookes called the "psychic force." They performed experiments designed to study the mysterious powers of the mind being exhibited by the prominent mediums and spiritualists of the day. The physicists thought that psychic forces might be connected with the electromagnetic "aether waves" that had just been discovered and were being put to amazing use. If wireless telegraphy was possible, why not wireless telepathy? This was a reasonable question at the time. However, while wireless telegraphy thrived, wireless telepathy made no progress in the full century of uncorroborated experiments in parapsychology that ensued.[7]

Conventional medicine follows conventional biology, conventional chemistry, and conventional physics in treating the material human body. It views the body as a complex, nonlinear system assembled from the same atoms and molecules that constitute (presumably) nonliving objects such as computers and automobiles. Medical doctors are in some sense glorified mechanics who repair broken parts in the human machine. Indeed, any stay in the hospital reinforces this image, as you are hooked to devices that measure blood pressure, temperature, oxygen saturation, and many other physical parameters. You are almost always treated with drugs that are designed to alter your body's chemistry. You usually get better—every time but one. Even so, you tend to view the whole experience as rather dehumanizing.

No surprise, then, that alternative practitioners find many eager patients when they announce that they go beyond materialism and mechanism and treat the really important part of the human system—the vital substance of life itself. People's religious sensibilities and images of self-worth are greatly mollified when they are told that they are far more than an assemblage of atoms—that they possess a living field that is linked to both God and the cosmos. Furthermore, the desperately ill will quite naturally seek out hope wherever they can find it.

Consequently, a huge market exists for therapists who claim that they can succeed where medical science fails. Studies conducted so far indicate that they do no better than a placebo, despite personal testimonies and media reports to the contrary. Even conventional physicians have been led to believe that some of these techniques, such as acupuncture, have merit, despite the fact that the data do not support this conclusion. Scientific medicine, by contrast, does immeasurably better than the alternatives, although many patients use both types of therapies and often misplace the credit for their cures.[8]

Many more studies looking at nonscientific methods can be expected in the near future, since the Center for Alternative Medicine within the U.S. National Institutes for Health received a huge increase in funding in 2000. Elisabeth Targ's deeply flawed distant healing studies, discussed in chapter 9, were funded by this agency. However, alternative practitioners have not always cooperated with scientists seeking to study their methods, claiming they go beyond science and so are not amenable to scientific tests.

Not all the proposed alternative therapies involve extraordinary phenomena, but most do. For example, although often conflated with "spiritual" healing, some ancient herbal remedies can reasonably be expected to contain some efficacious drug discovered through years of trial and error.[9] On the other hand, if homeopathic remedies have any effect at all, good or bad, then much of the body of knowledge in physics, chemistry, and biology, accumulated for over a century, would have to be discarded. According to the atomic theory of matter that is fundamental in these fields, a homeopathic dosage is so diluted that it is highly unlikely to contain even a single atom of the supposed active substance. I expect that we will see reports out of the center which claim evidence for the benefit of extraordinary therapies like homeopathy, published at the dubious 5 percent statistical significance level. Indeed, I predict that every twentieth experiment conducted will be so reported, while the other nineteen with negative results will go directly to the file drawer.

UNIFIED BIOFIELD THEORY

The hypothetical vital force is often referred to these days as the *bioenergetic field*. Touch therapists, acupuncturists, chiropractors, and many other alternative practitioners tell us that they can effect cures for many ills by "manipulating" this field, thereby bringing the body's "living energies" into balance. Presumably, unbalanced energies are unhealthy.

The use of "bioenergetic" in this context is ambiguous. This term is applied in conventional biochemistry to refer to the readily measurable exchanges of energy within organisms, and between them and their environment, which occur by normal physical and chemical processes. This is not, however, what the new vitalists have in mind. They imagine the bioenergetic field as a holistic living force that goes beyond reductionist physics and chemistry to the higher plane of spirit.

By "holistic" here, I am not referring to trivial homilies such as the need to treat the patient as a whole person and recognize that many factors, such as the psychological, emotional, and social, contribute to well-being, along with the health of the physical body. While unremarkable remedies of this sort are offered by those who claim to practice holistic medicine, they imply that something much more profound, something beyond reductionist materialism, is at work. But treating the whole person does not contradict any reductionist, materialistic principles. Neither does the fact that the parts of a physical system interact with one another—as long as they do so no faster than the speed of light. When data are presented showing that prayer or meditation lowers blood pressure or produces other beneficial physiological effects, this cannot be taken as evidence for the reality of spirit. The brain is wired to the rest of the body, so why should it not have an effect on the body's health? For example, it exerts enormous control over the manufacture of thousands of chemical substances by glands throughout the body.

Reductionism is not about a universe of isolated objects. It is about a universe of localized objects interacting with one another no faster than the speed of light. The holism that goes beyond reductionism implies a universe of objects interacting simultaneously, at infinite speed and so strongly that none of the pieces can ever be treated separately. In this nontrivial form of holism, all humans are connected to one another and to the cosmos as part of one inseparable unity. This concept enters into the discussion of bioenergetic fields, in which that field is imagined as some aether that pervades the universe and acts instantaneously over all of space. Quantum mechanics is vaguely presented as the agency for all this interconnectedness.

Therapeutic touch and other forms of energy therapies are now widely practiced within the nursing community.[10] These are based on a vague, unfounded theoretical system called "the science of unitary human beings," proposed by Martha Rogers.[11] According to Rogers, "Energy fields are postulated to constitute the fundamental unit of the living and nonliving." The field is "a unifying concept and energy signifies the dynamic nature of the field. Energy fields are infinite and paradimensional; they are in continuous motion."[12]

The exact nature of the bioenergetic field is not specified, in Rogers or the other literature on energy healing. On the one hand, the biofield is usually identified with the classical electromagnetic field. On the second hand, it is confused with quantum fields or wave functions. Yet it is something "spiritual," which would make it none of the other entities mentioned. An unfamiliarity with elementary physics is evident in much of what is written by the promoters of energy therapies. For example, Joanne Stefanatos writes, "The principles of energy medicine originate in quantum physics. Bioenergetic medicine is the study of human and animal bodies as dynamic electromagnetic fields existing in an electromagnetic environment."[13] Which is it? Quantum physics or dynamical electromagnetic fields? They are not the same thing.

AURAS AND DISCHARGES

Perhaps the most specific model for the bioenergetic field is some unrecognized form of electromagnetism. Advocates note that measurable electromagnetic waves are emitted by humans. This is often conflated with spirit and with quantum fields, but these electromagnetic waves are well understood and nothing extraordinary.

In the *Journal of Advanced Nursing*, Elissa Patterson relates "spiritual healing" to the belief that "we are all part of the natural harmonious energy of the universe." She then explains that within this universal energy field is a human energy field that is "intimately involved with human life, often called the 'aura.'"[14]

Some self-described psychics have said that they can "see" a human aura that they associate with a person's living energy field. This boast has not been substantiated empirically.[15] Indeed, humans do have auras, but these are not visible to the human eye. They must be photographed with infrared-sensitive film or viewed through special infrared glasses now commonly used by police and the military.

The infrared aura can be trivially identified with "blackbody" electromagnetic radiation. Everyday objects that reflect very little light, like you and me, will appear black, or unseen, in the dark. These bodies, however, emit invisible but detectable infrared light that is the result of the random thermal motions of the charged particles in the body. The wavelength spectrum has a characteristic smooth shape and is completely specified by the body's absolute temperature according to an equation that has been known for over a hundred years. As that temperature rises, the spectrum moves into the visible. The Sun, for example, radiates largely as a "blackbody" of temperature 6,000 degrees Kelvin, with a broad peak at the center of the visible electromagnetic spectrum, the color we perceive as yellow.[16] With their much lower body temperatures, humans radiate mostly in the infrared region of the spectrum that is invisible to the naked eye but easily seen with infrared detection equipment, such as that used by the military for night operations.

Surely, blackbody radiation is not a candidate for the bioenergetic field, for then even the cosmic microwave background, 2.7 degrees Kelvin radiation left over from the big bang, would be "alive." Blackbody radiation lacks any of the complexity we associate with life. It is as featureless as it can be and still be consistent with the conservation laws of physics. Any fanciful shapes seen in photographed auras emanating from humans can be attributed to optical and photographic effects, uncorrelated with any property of the body that one might identify as "live" rather than "dead," and the tendency for people to see patterns where none exist.

The inability of the nineteenth-century wave theory of light to explain the blackbody spectrum led, in 1900, to Max Planck's conjecture that light comes in bundles of energy called "quanta," thus triggering the quantum revolution. These quanta are now recognized as material *photons*. It is somewhat ironic that holists find such comfort in quantum mechanics, which replaced ethereal waves with material particles.

Stefanatos tells us that the "electromagnetic fields (EMF) emanating from bacteria, viruses, and toxic substances affect the cells of the body and weaken its constitution."[17] So the vital force is identified with electromagnetic fields and said to be the cause of disease. Actually, chemical forces are electromagnetic in nature, and chemistry is involved in disease, but this is not what is being implied here. Somehow, the life energies of the body are supposed to be balanced by bioenergetic therapies. "No antibiotic or drug, no matter how powerful, will save an animal if the vital force of healing is suppressed or lacking."[18] So health or sickness is determined by which side wins the battle between good and bad electromagnetic waves in the body.

Now it would seem that all these effects of electromagnetic fields in living things would be easily detectable, given the great precision with which electromagnetic phenomena can be measured in the laboratory. Physicists have measured the magnetic dipole moment of the electron (a measure of the strength of the electron's magnetic field) to one part in 10 billion and calculated it with the same accuracy. They surely should be able to detect any electromagnetic effects in the body powerful enough to move atoms around or do whatever happens in causing or curing disease. But neither physics nor any other science has seen anything that demands we go beyond well-established theories that contain no bioenergetic elements. No elementary particle or field has been found that is uniquely biological. None is even hinted at in our most powerful microscopes and other high-tech medical instruments. Maybe they are there at some undetectable level, but how do energy therapists know this? They do not show any new data beyond anecdotal testimonials.

Besides the infrared blackbody radiation already mentioned, electromagnetic waves at other frequencies are detected from the brain and other organs. As mentioned, these are often claimed as "evidence" for the bioenergetic field. In conventional medicine, they provide powerful diagnostic information. But "brain waves" show no special characteristics that differentiate them from the electromagnetic waves produced by moving charges in any electronic system. Indeed, they can be simulated with an electronic circuit. No marker has been found that distinguishes radiation from organisms as "live" rather than "dead."

Kirlian photography is often cited as evidence for the existence of fields distinctive to living things. For example, Patterson claims that "healers have the ability to see seven or more layers within an aura, each with its own color," and that these have been recorded using Kirlian photography.[19]

Semyon Davidovich Kirlian was an Armenian electrician who discovered, in 1937, that photographs of live objects placed in a pulsed high electromagnetic field will show remarkable surrounding " aura." In the typical Kirlian experiment, an object, such as a freshly cut leaf, is placed on a piece of photographic film that is electrically isolated from a flat aluminum electrode with a piece of dielectric material.[20] A pulsed high voltage is then applied between another electrode placed in contact with the object and the aluminum electrode. The film is then developed.

The resulting photographs indicate dynamic, changing patterns, with multicolored sparks, twinkles, and flares.[21] Dead objects do not produce such lively patterns! In the case of a leaf, the pattern is seen to gradually go away as the leaf dies, presumably emitting cries of agony during its death throes. Ostrander and Schroeder described what Kirlian and his wife observed: "As they watched, the leaf seemed to be dying before their very eyes, and the death was reflected in the picture of the energy impulses." The Kirlians reported that "we appeared to be seeing the very life activity of the leaf itself."[22]

As has been amply demonstrated, the Kirlian aura is nothing but *corona discharge*, reported as far back as 1777 and completely understood in terms of well-known physics. Controlled experiments have demonstrated that claimed effects, such as the diminishing aura of a dying leaf, are sensitively dependent on the amount of moisture present. As the leaf dies, it dries out, lowering its electrical conductivity. The same effect can been seen with a long-dead but initially wet piece of wood.[23]

Once again, like the infrared aura, we have a well-known electromagnetic phenomenon being marketed to an unwitting public, unfamiliar with basic physics, as "evidence" for a living force. It is nothing of the sort. Proponents of alternative medicine would have far fewer critics among conventional scientists if they did not resort to this kind of dishonesty and foolishness.

QUANTUM MYSTICISM

The word "quantum" appears in virtually everything written on alternative medicine. One can also find it in books and articles that seek a theoretical basis for ESP (extrasensory perception) and other psychic and mystical notions.[24] It is the vague magical incantation uttered in order to make all the inconsistencies, incoherencies, and incompatibilities of the proposed scheme disappear in a puff of smoke. Since quantum mechanics is weird, anything weird must be quantum mechanics.

Quantum mechanics is claimed as support for ESP and the related notion of mind-over-matter often called "psychokinesis."

These various postulated psychic phenomena are often simply referred to as *psi*, from the Greek letter Ψ. This happens to also be the symbol conventionally used in quantum theory to represent the *wave function* that specifies the quantum state of a system, a convenient coincidence for those who claim to see a connection with psychic notions. The way the observer is entangled with the object being observed in quantum experiments is erroneously used to infer that human consciousness actually controls reality. As a consequence, we can all think ourselves into health and, indeed, immortality—if only we buy this book.[25] "Quantum healing" is based on a particularly misleading interpretation of quantum mechanics. Other interpretations exist that do not require that anything mystical is going on.[26]

The authority of Einstein is also frequently invoked in the literature on bioenergetic fields, and, once again, deceptively so. Stefantos says: "Based on Einstein's theories of quantum physics, these energetic concepts are being integrated into medicine for a comprehensive approach to disease diagnosis, prevention, and treatment."[27] Einstein's theories of quantum physics? Which theories are these? While Einstein contributed to the early development of quantum mechanics, especially with his 1905 photon theory, modern quantum mechanics is the product of a large group of physicists in the period 1900–1930. Max Planck, Niels Bohr, Louis de Broglie, Werner Heisenberg, Erwin Schrödinger, Wolfgang Pauli, Max Born, Pascual Jordan, and Paul Dirac each made contributions to quantum mechanics at least as important as Einstein's.

Referring to well-known but off-beat authorities such as Fritjof Capra and Ken Wilber, Stefanatos tells how "Einstein's quantum model replaced the Newtonian mechanistic model of humankind and the universe."[28] Thus holistic healing is associated with the rejection of classical Newtonian physics. Yet holistic healing retains many ideas from eighteenth- and nineteenth-century physics, in particular, etheric fields. Its proponents are blissfully unaware that these ideas, especially superluminal holism, have been rejected by modern physics as well.

Never mind that Einstein was not the inventor of quantum mechanics and objected strongly to its statistical character, saying famously, "God does not play dice." Never mind that electromagnetic fields were around well before quantum physics and it was Einstein himself who proposed that they are composed of reductionist particles, now called protons, rather than holistic fields. And never mind that Einstein did away with the aether, the medium that nineteenth-century physicists thought was doing the waving in an electromagnetic wave and a few others thought might also be doing the medium for "psychic waves." The bioenergetic field described in holistic literature seems to be confused with the aether, and this, in turn, confused with spiritual or psychical fields. Or perhaps no confusion is implied. They all seem to share at least one feature in common—nonexistence.

As the nineteenth century drew to a close, experiments by Albert Michelson and Edward Morley had failed to find evidence for the aether. This laid the foundation for Einstein's theory of relativity and his photon theory of light, both published in 1905. Electromagnetic radiation is now understood to be a fully material phenomenon. Photons have both inertial and gravitational mass (even though they have zero rest mass) and exhibit all the characteristics of material bodies. Electromagnetism is as material as breath, and an equally incredible candidate for the vital field.

However much it might be wished otherwise, the fact remains that no animating force unique to living organisms has ever been conclusively demonstrated to exist by any scientific experiment. Of course, evidence for a life force might someday be found, but this is not what is claimed in the literature that promotes much of alternative medicine. There you will see the strong assertion that scientific evidence currently exists for some entity beyond conventional matter and that this claim is supported by modern physical theory—especially quantum mechanics. Furthermore, the evidence presented in this literature is not in the form of data from our most powerful microscopes or particle accelerators, probing beyond existing frontiers. Rather, it resides in vague, imprecise, anecdotal claims of the alleged curative powers of traditional folk remedies and other nostrums. These claims simply do not warrant such a profound interpretation by even the most minimal application of scientific criteria. A bioenergetic field plays no role in the theory or practice of biology or scientific medicine. Vitalism and bioenergetic fields remain hypotheses not required by the data.

As already noted, most alternative medicine is based on claims that violate well-established scientific principles. Proponents of treatments that rest on the existence of a bioenergetic field, whether therapeutic touch or acupuncture, should be asked to meet the same criteria as anyone else who claims a phenomenon whose existence goes beyond established science. They have an enormous burden of proof, and they should not be allowed to avoid that burden.

THE SEARCH FOR PSYCHIC ENERGY

The second half of the nineteenth century saw the rise of *spiritualism*, a social phenomenon in which thousands of individuals, called mediums, claimed they could make contact with the world beyond. The whole medium frenzy began in 1848 with Kate and Margaret Fox, young sisters from Rochester, New York, whose messages from the spirits could be heard by listeners as audible rapping sounds. In 1850 they became headline attractions at P. T. Barnum's American Museum in Manhattan (where "a sucker is born every minute"). Years later, Margaret admitted that it was all a hoax, performed by the girls cracking their toe joints—exactly as suggested by skeptics from the very beginning.[29] By that time, however, spiritualism was in full swing and was not about to be stopped in its tracks.

The mediums who followed the lead of the Fox sisters typically staged the familiar séance in which several people sit around a table in a darkened room. They would experience many unusual happenings after the lights were out: strange noises, flashes of light, the table vibrating and rising into the air. The medium might even seem to fly above the table. In one séance, several eyewitnesses swore that famed medium Daniel Dunglas Home floated out one window of the séance room and into another. In their most popular feat, mediums acted as intermediaries for bereaved people who wished to contact departed family members.

While séances are rare today, modern mediums, now called channelers or psychics, still talk to the dead. Several have recently made fortunes for themselves from television appearances and books by assuring the grief-stricken that their loved ones are happy in Heaven. Unfortunately, they seem to be able to only extract trivial homilies from the spirits: the living need to learn to love one another, eat less red meat, and so on. They never report useful information, such as the winner of the next Kentucky Derby or the location of buried treasure. Of course, it is always possible that spirits lack access to worldly information. Still, if the modern mediums came up with some verifiable prediction, that would give credibility to a spirit world. The medium never provides any information that could not be discovered by the standard techniques used by mentalists and fortune-tellers for thousands of years.

The ostensible spiritual powers of the original nineteenth-century mediums became the subject of numerous scientific tests. In many cases, the mediums were revealed as charlatans who simply used common magician tricks. Many had been professional stage magicians before finding more lucrative employment. The famous escape artist Harry Houdini showed little patience with their dishonest application of an ancient art and exposed many of the tricks used.

In a few cases, though, mediums seemed to pass all the tests that scientists and magicians could throw at them. This convinced some of the more gullible scientists that the mediums were legitimate. I have already mentioned two prominent physicists, William Crookes and Oliver Lodge, who performed their own experiments and became convinced of the reality of psychic or spiritual forces. Indeed, in 1871 Crookes examined Kate Fox and declared that her spirit rappings were genuine, true objective occurrences. Both Crookes and Lodge were later shown to have been fooled by their subjects on a number of occasions. It should be noted that both had strong religious beliefs that may have affected their objectivity.[30]

In 1882 the Society for Psychical Research was organized in England, to be followed soon after by the American Society for Psychical Research. These remain in existence, and their journals are still being published today.

At the time, skeptics pointed to a number of flaws in the way psychic research was being conducted. Most of the investigations simply collected anecdotal reports about haunted houses, ghostly apparitions, precognitive dreams, and other strange happenings that could not be properly confirmed or analyzed. Perfectly natural explanations for these events were proposed by skeptics: chance, hallucinations, hoaxes, trickery, and any number of other nonsupernatural possibilities. Lacking sufficient data, however, the skeptics could not always prove an alternate hypotheses. Logically, that provided no basis to assume the paranormal events were true. But some people would argue that way, expecting skeptics to bear the burden of disproving the least-parsimonious explanations of the phenomena, that supernatural events had taken place.

With anecdotal stories alone as the source of data, when reliable, independent information is not available, the best one can do is propose alternative, plausible scenarios that are simpler and more economical than any paranormal explanation. Often the scenario begins with the suggestion that the source of the anecdote is lying, which rubs many people the wrong way. We, in the United States at least, are used to assuming someone is innocent until proven guilty. When one is making an extraordinary claim, however, he may be assumed to be guilty until he proves himself innocent. This is another feature of scientific method that many people find offensive because it clashes with their sensibilities. Yet science would not progress if it was forced to study every unsubstantiated proposal no matter how bizarre.

Even many scientists who were sympathetic to the study of psychic forces agreed that something as extraordinary as psychic phenomena would never be confirmed by anecdotal evidence alone. Thus Crookes, Lodge, and others who followed attempted to assert a modicum of control in their experiments. But they still made the fatal mistake of allowing the medium to dictate the terms of the experiment, which were usually also conducted on the medium's own turf, in the typical darkened room. Being upright men themselves, they could not imagine so much cheating out there in the real world—that someone would actually resort to trickery in a scientific investigation. The cathode rays and electromagnetic waves that Crookes and Lodge studied in their normal research did not lie to them, so it was unthinkable that any human object of experimentation would do so.

It was not until the 1930s, with the work of another upright religious man, Joseph Banks Rhine at Duke University, that psychic powers began to be tested in experiments in which the scientists retained control and seemingly more careful procedures were put in place. Even these turned out to be insufficient to rule out the possibility of deception. Rhine's typical experiment involved one person, the "sender," choosing an item randomly, often a card from a shuffled deck of five "ESP cards," which have since become very familiar. These display a simple cross, square, circle, star, or three wavy lines. The sender concentrates on the image of the selected card while a "percipient" then tries to read the sender's mind.

Although Rhine claimed to find statistically significant evidence for ESP, a term he coined, the bulk of the scientific community was not convinced. Skeptics found much to criticize in Rhine's protocols. For example, someone showed that, for at least one set, the ESP cards were so thin that the image was visible from the back! Also, until Rhine started to tighten up his procedures, the sender and recipient were usually close enough to one another that sensory clues remained quite possible.

Unable to get published in the regular scientific journals, Rhine started his own peer-reviewed journal, in which he personally selected the "peers" who refereed the submissions. Unsurprisingly, they rarely rejected a Rhine paper. Ultimately, however, sometimes years later, flaws were found in all of Rhine's published experiments, including the two or three that he personally labeled the best. Rhine's subjects were mostly students who were paid to participate, and those who were more successful were retained on the assumption that they possessed special mental talents. They apparently did, but not exactly the psychic talent Rhine had assumed. Unwittingly, Rhine provided his subjects, and his research assistants, with a strong incentive to provide the results he expected. Indeed, outright faking of data by a co-investigator was uncovered in the Rhine laboratory and in another laboratory in England where similar work was being done, although Rhine himself was not implicated.[31]

Rhine maintained his reputation as an honest scientist. Unfortunately, like Crookes and Lodge, he was gullible and overly committed to a specific conclusion of his research. Rhine knew full well that if he were to find the first convincing evidence for ESP, and perhaps, by implication, the existence of a spirit world, he would become one of the most famous men in history. But as we have already seen, science can become pathological when a scientist's dreams of glory cloud what the data are saying.

Rhine founded the field of parapsychology that continues today in a small number of privately funded institutes, only a handful of which are associated with universities. (Duke has long since dropped the Rhine program, although an independent private institute remains nearby). From the time of Rhine, the history of parapsychology has been marked by breathless media reports of new, convincing evidence for psychic phenomena, followed by the quiet failure of these observations to hold up under further scrutiny.

Parapsychologists have convinced no one but themselves and a credulous public who wants to believe in exceptional human powers. Again, a good portion of the debate pivots on the interpretation of statistics. I have already mentioned that some parapsychologists ask to be judged by the same criteria used in psychology and medicine, in which a result is published when one out of twenty results can be due to chance alone. I was present at a talk by statistics professor and psi-believer Jessica Utts when she argued that the statistical evidence for psychic phenomena is better than that for the studies showing that an aspirin a day helps prevent heart attacks. If we believe the aspirin result, she asked, then why don't we believe the psi result?

The difference, which Utts does not accept, is between extraordinary and ordinary claims. The claim that aspirin helps prevent heart attacks is an ordinary one that may be wrong but does not violate any known medical principles and, in fact, has a reasonable theoretical basis. Furthermore, if it does no harm to take an aspirin a day, and it may help, then why not publish this fact and let patients and their physicians decide whether to use this treatment option? My only problem with this is that patients and physicians are not always aware that a published claim has a good chance of being wrong, given the willingness of medical journals to accept five false conclusions out of every one hundred. Physicians and patients are routinely misled into believing that the study would not have been published if the evidence were weak.

The existence of psychic phenomena, on the other hand, is an extraordinary claim and, as such, sits in a different class of scientific questions than the value of aspirin in preventing heart disease. Five in one hundred false reports is far too loose for the type of extraordinary claim being made here, that the mind has the power to breach the laws of nature. We simply cannot take every twentieth experiment that comes along with a normal statistical fluctuation and use it to discard otherwise well established scientific knowledge.

MIND-MACHINE MACHINATIONS

One experiment that has been running continuously since 1979 claims evidence far exceeding any reasonable threshold of statistical significance for the power of the mind to breach the laws of nature. This investigation has been carried out by the Princeton Engineering Anomalies Research group (PEAR) under the direction of Robert G. Jahn, former dean of the School of Engineering and Applied Science at Princeton.

Jahn's experiment is simple, inexpensive, and easily repro-
ducible—though time-consuming. Subjects are asked to mentally
control the output of a random number generator and make it
deviate from chance. PEAR has accumulated thousands of trials
using hundreds of subjects and, as of 1997, claimed an effect size
of one part in ten thousand. That is, the subject is able to affect the
outcome of the random number generator once every ten thousand times, on average. The probability for this observed effect or
a greater one being the result of chance is presented as exceeding
10 trillion to one.[32] Astronomer and statistics expert William Jefferys (private communication), who has studied the PEAR results,
has provided me with this precise formulation of the meaning of
these results, which I paraphrased in chapter 9: "If we assume that
the null hypothesis is exactly true, and repeat the whole experiment many times (including all N trials for each repetition), then
in the long run we will obtain effects this great or greater in only
one in 10 trillion trials."

Interestingly, the random numbers are sometimes generated
before the subject is given a chance to mentally affect their outcome.
Thus Jahn and collaborators imply they not only have evidence for
mind over matter (they refer to it as a "mind-machine interaction"), they also imply they have evidence that the mind can act
backward in time. In an article and book written by Jahn and his
chief assistant, Brenda Dunne, quantum mechanics is presented as
providing a scientific basis for the phenomenon.[33]

In my book *Timeless Reality*, I discuss in detail how time
reversibility seems to be evident at the quantum scale. However, as
I show, it does not necessarily follow, and is indeed very unlikely,
either that objects as large as humans can travel back in time, or
that thoughts can influence events in the past. In any case, as
always, let the data decide.

The PEAR team insists that it has answered the numerous criticisms made by skeptics of their earlier results.[34] The statistical significance of the first set of effects reported was almost entirely
attributable to one subject, who also happened to be one of the
experimenters. They now allege that significant effects are broadly

distributed among many subjects. However, their latest results still show some puzzling characteristics that have not been satisfactorily explained.

For example, in any experiment it is important to have a control against which to measure the deviations that signal some new effect. In the case of PEAR, this is called the "baseline." The baseline is formed by data that are taken when the subjects are not trying to affect the outcome of the random number generator. A careful analysis of this baseline data indicates it is too good to be true, since it does not display the full range of deviations expected from statistical fluctuations.[35] This is explained away by suggesting that the subjects, either consciously or subconsciously, may have been trying to produce a baseline better than chance. This is reminiscent of other explanations given by parapsychologists for results that do not work out as expected, such as the famed "observer effect" in which the presence of skeptics in the laboratory induces bad vibrations that neutralize the psychic force. This might be called the "believer effect" in which no observation is allowed to imply that the wished-for phenomenon may not be present.

Despite constructive suggestions from skeptics (yes, skeptics do try to be helpful), the PEAR group has not departed from its quite unconventional procedure of not carefully monitoring its subjects. This protocol is severely frowned upon by most scientists, including some parapsychologists, because it does not sufficiently rule out the possibility of cheating. The same can be said of any other systematic bias in the experiment. PEAR has not changed another very dubious procedure, mentioned above, in which experimenters also act as subjects. As any scientist who deals with small effects will tell you, a small systematic error can have an enormous impact on the calculated statistical significance. For example, to produce the PEAR effect by cheating, only one person out of a hundred would have to cheat, on average, once every hundred times. PEAR claims this is impossible, but since they do not watch their subjects, how do they know?

In yet another highly questionable procedure, subjects are given immediate feedback on their attempts to control the random number sequence and are allowed to modify certain aspects of the experiment. This can also lead to biased data, as some computer simulations have suggested. It is not clear that the rather conventional statistical methods used by the experimenters apply for such a tiny effect accumulated over such a huge number of trials under the conditions used.

Like many of the other arguments I make in this book, I cannot prove that any of these criticisms explain the reported observations. But, again, that is not my burden. Over the years, PEAR experimenters have conducted many analyses and tests that they insist rule out any of the explanations proposed by skeptics. However, they have steadfastly refused to significantly modify their protocols to meet the main objections. They are like chemists who do not clean their test tubes, even though the sink is right over there in the corner of the lab, insisting that they know how to subtract the dirt from their sample theoretically. One can only speculate why PEAR has not just bitten the bullet and said: Okay, we are sure we have already eliminated all sources of error. Nevertheless, from now on we will: (1) not count any data taken with experimenters as subjects, (2) not allow the subjects to modify the experiment in any other way, and (3) monitor the subjects at all times. Then let us see what happens.

It must be noted that despite the basic simplicity of the PEAR experiment, the results have not been independently replicated with a result of the same effect size in the more than two decades the experiment has run. Perhaps this is because of the length of time needed to do the necessarily large number of trials. But one would still expect that such a profound claim would be widely checked were it taken seriously. Earlier work on random number generators by Helmut Schmidt claimed an effect, but with a much greater effect size than PEAR.[36] This is often incorrectly touted as a replication. However, two experiments cannot be said to replicate one another when their results do not agree quantitatively. In this case, PEAR and Schmidt should be regarded as *failing* to confirm each other.

Over many years, other types of experiments have attempted

to detect the ability of thoughts to control physical events.[37] Of particular note are those experiments where the humans try to mentally modify very delicate quantum phenomena.[38] The mind-matter interaction is usually postulated to happen at the quantum level, so that should be the best place to look. However, none of these mind-over-matter experiments have discovered any significant effects. Thus, PEAR stands alone.

If the PEAR experimental results are correct, they imply that some humans can mentally affect an otherwise random outcome once out of ten thousand times. Even this interpretation is questionable, since they find no difference between their data taken with (presumably) true random numbers generated by quantum noise and pseudorandom numbers generated by computer algorithms. Practical applications are dubious, although the group is now involved in developing products that they hope to sell. Unfortunately, a one-in-ten-thousand edge is too small to beat the house in gambling casinos and, to my knowledge, no PEAR subjects, investigators, or customers have become fabulously wealthy.

METANALYSIS AND THE FILEDRAWER EFFECT

In the natural sciences, an extraordinary phenomenon is normally not considered to be even tentatively established until it has been observed, in precisely the same form, in two or more independent experiments in which each experiment stands alone as being statistically significant. It is safe to say that this condition has not yet been achieved for psi or spiritual phenomena, despite experiments going back to the 1850s. Any other phenomenon that had failed to be confirmed after all this time would have been long abandoned as a lost cause. But since scientific evidence for psi would confirm the belief of so much of the human race that they possess mental or spiritual powers transcending the limitations of matter, the search goes on.

While many parapsychologists admit that the existence of psi is still not conclusively demonstrated, a few have insisted that the evidence in its favor is now overwhelming. Since they are not able to make this assertion based on conventional scientific criteria, they invent other criteria. Current claims rest on a dubious procedure called *metanalysis* in which the statistically insignificant results of many experiments are combined as if they were a single, controlled experiment. None of the individual experiments have anything approaching the physics threshold for the publication of an extraordinary claim, 10,000-to-1 odds against chance. Most do not even meet the weak 20-to-1 threshold which is, at best, barely adequate for ordinary claims. PEAR hangs alone in its tree with no other PEARs to metanalyze.

Where several similar experiments are available, the metanalyzed probability for the whole package resulting from chance is estimated using statistical techniques. If the combined result is statistically significant, then the phenomenon is regarded as confirmed. I cannot think of a single example of a new phenomenon in science that has been established by metanalysis.

The most prominent proponent of using metanalysis to demonstrate the reality of psi is Dean Radin, whose 1997 book, *The Conscious Universe*, is subtitled *The Scientific Truth of Psychic Phenomena*.[39] Radin asserts that when one looks at the aggregate of data collected over time, one must conclude that psychic phenomena are scientifically validated.

For example, Radin notes that 186 ESP card tests involving 4 million trials were published worldwide from 1882 to 1939. He takes these results at face value, downplaying any possibilities of cheating or other plausible explanations that skeptics have been able to uncover in virtually every case in which sufficient information about procedures has been made available. Making the disputable assumption that the data are all trustworthy, Radin claims that the odds against chance are more than a billion trillion to one.

Radin is aware of the filedrawer effect in which only positive results tend to get reported and negative ones left in the filing cabinet. This obviously can greatly bias any analysis of combined results, and Radin cannot ignore this as blithely as he ignores other possible, nonparanormal explanations of the data. Even the most fervent parapsychologists recognize this problem. Metanalysis incorporates a procedure for taking the filedrawer effect into account. Radin says it shows that more than 3,300 unpublished, unsuccessful reports would be needed for each published report in order to "nullify" the statistical significance of psi.

In his review of Radin's book for the journal *Nature*, statistics professor I. J. Good disputes this calculation, calling it "a gross overestimate." He estimates that the number of unpublished, unsuccessful reports needed to account for the results by the file-drawer effect should be reduced to fifteen or less.[40] How could two metanalyses result in such a wide discrepancy? Somebody is doing something wrong, and in this case it is clearly Radin. He has not performed the filedrawer analysis correctly.

Douglas Stokes is a specialist in statistical analysis who is very sympathetic to the psi movement. In his book *The Nature of Mind*, Stokes considers the wide range of reports of psychic phenomena, from the anecdotal to the experimental. Although he concludes that psi has not been scientifically demonstrated, he still wants to believe in it based on the "compelling stories" he has heard and his own personal "spontaneous psi experiences."[41]

Stokes is thus no closed-minded skeptic, no dogmatic nonbeliever. Nevertheless, he has published in *Skeptical Inquirer*. Specifically, he has described in that publication a fundamental error that Radin and others have made in their calculations.[42]

The filedrawer effect is not confined to metanalysis but applies to single experiments as well. Stokes examined the filedrawer analyses in several published psi experiments. He found that the authors had grossly overestimated the number of null experiments that would have had to remain unreported to obtain the combined positive results reported. He looked specifically at an experiment by Alan Vaughn and Jack Houck involving ESP-guessing questionnaires sent to them by 12 subjects, each significant at the 5 percent level.[43] The authors claimed that over 33,000 subjects would have had to be tested to produce the reported effect as a statistical fluctuation. Since they did not send out anywhere near that number of tests, they conclude that the net effect was real.

To test this calculation, Stokes simulated the experiment on a computer, which the authors also could have easily done, and indeed should have done. This is called a "Monte Carlo analysis." I spent a good portion of my research career in particle physics doing such analyses, in which you try to learn all the possible sources of error in an experiment, statistical and systematic, by "doing the experiment in the computer." This procedure is simple, straightforward, and does not rely on any problematic statistical techniques or packaged programs that the users apply blindly. One obvious fact that strikes me when reading many of the published papers on psi and religious scientific claims is the authors' often complete ignorance of statistics.

As an example, Stokes generated random data for 30 subjects and selected out the 12 highest scores, all of which had p-values of 1 percent or better. This left only 18 in the file-drawer, not 33,000 as Vaughn and Houck claimed were needed. The experiment was designed so that the subjects knew their scores before mailing them in. One can easily imagine 18 with low scores not bothering to report.

I regard it as especially significant that this careful analysis was done by a psi sympathizer. Critics of paranormal claims are often accused of being closed-minded, dogmatic worshippers of the "religion of scientism." This is especially the case for those, like myself, who write for the *Skeptical Inquirer,* a publication of the Committee for the Scientific Investigation into Claims of the Paranormal (CSICOP). We are often derisively called the "psicops," implying that we are engaged in a prosecution of psi as if it were some kind of crime.

Of course, we reject this offensive charge. We are only devoted to science as the best means humans have developed, so far, for arriving at an approximation to the truth about objective reality—whatever that truth may be. We are not closed-minded against psi, religion, alternative medicine, or any paranormal claims, nor prejudiced against any individual adherent. Show us the evidence and we will consider it. However, we steadfastly insist on the same rules that we would apply to claims for a new particle or a new drug. In particular, we refuse to agree to adopt new criteria, such as proposed by Radin, solely for the benefit of researchers in a field of study that cannot seem to get significant positive results any other way.

Radin, as others before, has made much of the metanalyses performed on so-called ganzfeld experiments. In these experiments, subjects are placed in a sensory-deprived state, with halves of Ping-Pong balls taped to their eyes and white noise in the background. They attempt to use their psychic powers to describe pictures that are being viewed by senders in a distant room. The "hits" and "misses" are later judged by disinterested individuals not part of the experimental team.

In 1985 Charles Honorton published a metanalysis of 28 ganzfeld experiments. He estimated that the averaged results of 423 unreported studies with insignificant results would have been needed to account for the results of the 28 reported experiments as a filedrawer effect.[44] He asserted that the odds against the aggregate result being the result of chance was over a billion to one. Over 400 unreported studies certainly seemed unlikely, since parapsychologists would probably be aware of that many ganzfeld experiments being done around the world.

However, Stokes discovered a fundamental mistake in this metanalysis. Honorton assumed that the unreported studies gave results expected from chance. In fact, since those with positive results were selected out, the remainder must average *below* chance. Correcting for this, Stokes found he could get the same results with 62 unpublished reports, a number that could very well have gone unnoticed.

Why this vast difference? In a detailed analysis of the filedrawer problem, Jeffrey Scargle has shown that a widely used technique called "fail safe file drawer" is deeply flawed because it assumes that the experiments in the file-drawer are unbiased.[45] In fact, they are biased, by definition. As Stokes showed in the specific case above, this technique leads to a grossly wrong overestimate of the size of the file-drawer.

And so we see why metanalysis is not utilized in those sciences that seek out remarkable new phenomena. While, in principle, it takes the filedrawer effect into account, it does so in a way that is far too sensitive to the slightest mistakes in analysis and too unreliable to be used as the only basis for an extraordinary claim.

RELIGIOUS EXPERIENCES AND THE BRAIN

Yet another place where science interacts with religion is in the interpretation of so-called religious or mystical experiences. These have been reported throughout history and across cultures. While details of such experiences vary greatly, depending on culture, the subject generally reports intense feelings of separation from the physical world and entry into what seems to them to be a spiritual state in which they feel at one with God and the cosmos.

Great religious leaders of the past, for example Buddha, Jesus, Paul, and Mohammed, along with many lesser figures, are said to have had such experiences.[46] These were taken by their followers as evidence of the special sacredness of their masters, and the messages obtained during these experiences were adopted as divine revelations. Many of the otherwise ordinary people who have had such experiences report how their lives were changed by what seems to them as irrefutable evidence for a transcendent reality. As mentioned in the preface, this ancient method of acquiring knowledge independent from the senses is called *mythos*. But how does the mystical method of *mythos* compete with the logical-empirical method of *logos* when it comes to delivering reliable information about the reality outside our heads?

One very common variety of mystical encounter is the near-death experience (NDE), brought to public attention in the mid-1970s by the best-seller *Life After Life* by Raymond Moody.[47] NDEs have come to the fore in recent times because of advanced medical techniques in which a person may be resuscitated after his or her heart and breathing have stopped. Presumably these people are not dead, since their brains remain active, and they often report a rather pleasant experience. Moody described the typical event in which a person hears himself being pronounced dead and traverses a long, dark tunnel at the end of which lies an encounter with a "being of light." Although he feels joy and peace, he returns to his body and life.[48]

In the 1980s, Kenneth Ring surveyed 102 people who had come close to death and found that half had a "core experience," which Ring described as a sequence of five stages: peace, body separation, entering darkness like a tunnel, seeing the light, and entering the light. More people experienced the earlier stages than the later stages.[49]

No one reporting such an experience has exhibited the so-called flat line on an electroencephalograph that is, somewhat controversially, called "brain death." It should also be noted that similar experiences have been reported by people who were not close to death.[50]

Nevertheless, the experiences are very real, according to the people who report them, and not simply the same as a pleasant dream. These people insist that they did not imagine what happened, and many observers regard this sincere testimony as evidence enough for a spirit realm and the existence of life after death. Still, no one has ventured beyond the light at the end of the tunnel and returned with a description of what is there.

A similar transcendent reality is claimed for the out-of-body experience (OBE) in which someone lying on a bed imagines herself floating above her body. In New Age jargon, this is called *astral projection*, in which the spirit is supposed to move to a higher plane of existence. These are even more common than NDEs and can occur under normal conditions. Perhaps as many as 50 percent of marijuana users report having such an experience sometime in their lives.[51]

Scientists are learning much about what goes on in the brain during a mystical experience. While something physical is no doubt happening, great doubt exists as to whether anything more than a purely material process is also involved. Similar phenomena can be induced in any number of physical and chemical ways: with psychedelic drugs, electrical and magnetic stimulation of the brain, and migraine headaches. They can be symptomatic of diseases such as epilepsy and schizophrenia. Pilots undergoing high-acceleration training in a centrifuge also report effects similar to the NDE as their brains are deprived of oxygen. Thus cerebral anoxia or hypoxia can account for much of what is reported, and the NDE dark tunnel with a light at the end has a plausible physiological explanation.[52]

Of special note is the work of Canadian neuroscientist Michael Persinger and his collaborators. Persinger has discovered that a weak, frequency-modulated magnetic field applied to the temporal lobe of the brain will cause most people to feel a spectral presence in the room with them. Here is the summary of a recent experiment:

> To test the hypothesis that experiences of apparitional phenomena with accompanying fear can be simulated within the laboratory, a forty-five-year-old journalist and professional musician who had experienced a classic haunt four years previously was exposed to 1 microTesla, complex, transcerebral magnetic fields. Within ten minutes after exposure to a frequency-modulated pattern applied over the right hemisphere, the man reported "rushes of fear" that culminated in the experience of an apparition. Concurrent electroencephalographic measurements showed conspicuous one-second-to-two-second paroxysmal complex spikes (15 Hz) that accompanied the reports of fear. A second magnetic field pattern, applied bilaterally through the brain, was associated with pleasant experiences. The subject concluded that the synthetic experience of the apparition was very similar to the one experienced in the natural setting. The results of this experiment suggest that controlled simulation of these pervasive phenomena within the laboratory is possible and that this experimental protocol may help discern the physical stimuli that evoke their occurrence in nature.[53]

James Austin, author of *Zen and the Brain*,[54] has combined his own experiences with Zen mediation and his work as a neuroscientist. He suggests that during meditation the circuits in the brain that normally deal with monitoring the environment go quiet and that other circuits associated with self-awareness disengage. This suspension of conscious orienting of oneself to one's surroundings then leads to the impression of otherworldliness that is reported.

This interpretation is supported by brain imaging techniques that have pinpointed specific areas of enhanced activity and inactivity in the brain during a mystical experience. In *Why God Won't Go Away*, neurologist Andrew Newberg tells about the experiments he and his late colleague, Eugene d'Aquili, performed on eight Buddhists engaged in Tibetan-style meditation and several Franciscan nuns in prayer. Although d'Aquili is listed as a co-author, the book was written after his death, and he is referred to throughout in the third person. I will assume that the words in the book are Newberg's.

Newberg does not make clear if the "Tibetan meditators" were actual Tibetans or Westerners who use a Tibetan meditation technique. The only Tibetan meditator mentioned is named "Robert," and a 2000 *Newsweek* cover story (Sharon Begley again) seems to imply that he was a colleague of Newberg's, Michael Baime.[55] Also, for some reason, no control group of relaxed atheists or stoned bikers was included.

While both of the groups studied reported mystical experiences, it is important to note that these were quite different between groups and obviously affected by their particular belief systems. Robert reported a sense of "timelessness and infinity," as if he were "part of everything in existence."[56] The other Buddhists apparently had similar experiences, although these are not detailed in the book. The Catholics, on the other hand, "tended to describe this moment as a tangible sense of the closeness of God and a mingling with Him."[57] Neither group reported revelations about future events—risky predictions that could be used to objectively confirm a true otherworldly reality to the experience. All they could say was that it seemed real to them. Now, they can argue that that was not their intent, that this is not the purpose of mysticism. I am simply providing the reader with an example of the sort of information that would give their claims some scientific credibility, some indication that it was not just all in their heads.

The brain imaging was performed using a SPECT camera, the acronym standing for *single photon emission computed tomography*. The primary result for all meditators was decreased brain activity in what the authors dubbed the orientation association area (OAA) of the brain. Newberg explains that the function of this area is to "draw a sharp distinction between the individual and everything else, to sort out the you from the infinite not-you that makes up the rest of the universe."[58]

The drop in activity in the OAA presumably results from the decrease in the flow of incoming sensory information during meditation. Newberg surmises that without this information the OAA cannot find the boundary between "you and not-you." Thus, "the brain has no choice but to perceive that the self is endless and intimately interwoven with everyone and everything the mind senses."[59]

This association of specific, localized brain activity and the highly subjective nature of the reports strongly implies a purely internal, biological origin of the belief that arises from religious experiences. Newberg seems to affirm this when he writes:

> Evidence suggests that the deepest origins of religion are based in mystical experience, and that religions persist because the wiring of the human brain continues to provide believers with a range of unitary experiences that are often interpreted as assurances that God exists.[60]

He adds,

> It is unlikely that the neurological machinery of transcendence evolved specifically for spiritual reasons. Still, we believe that evolution has adapted this machinery, and has favored the religious brain because religious behaviors turn out to be good for us in profound and pragmatic ways.[61]

Newberg concludes at this point that "God cannot exist as a concept or as reality anyplace else but in your mind."[62] But elsewhere he seems to contradict himself by implying just the opposite:

After years of scientific study, and careful consideration of our results, Gene and I further believe that we saw evidence of a neurological process that has evolved to allow us humans to transcend material existence and acknowledge and connect with a deeper, more spiritual part of ourselves perceived of as an absolute, universal reality that connects us to all that is.[63]

Conventional thinking, according to Newberg, explains religious thinking as a cognitive process we dream up for comfort and protection. However,

A neurological approach . . . suggests that God is not the product of a cognitive, deductive process, but was instead "discovered" in a mystical or spiritual encounter made known to human consciousness through the transcendent machinery of the mind.[64]

And,

Science has surprised us and our research has left us with no choice but to conclude that the mystics may be on to something, that the mind's machinery of transcendence may in fact be a window through which we can glimpse the ultimate realness of something that is truly divine."[65]

So God is not just in the mind after all, Newberg says, but the mind detects him during a religious experience.

Of course, this is what religions have been claiming all along, and Newberg would have us think that he has produced some kind of scientific evidence for this in his "neurological approach." But nowhere does he tell us what that evidence is. The experiments he conducted, in fact, imply just what he said in the first place: Religious experiences, or their reasonable facsimile, are correlated with internal brain processes that have a perfectly natural, material explanation. They seem real to the people experiencing them, however, who are often convinced that they have linked up to a deeper reality. In some cases this reality is described as being "more real" than the normal awake state. But even a sympathetic investigator can go no further than to take their word for it. They bring back no information by which we can objectively confirm the existence of this reality, nothing that was not in their heads all along.

Again I must emphasize that I am not claiming that the evidence rules out every possible supernatural interpretation that might be given, or that it proves the fully natural explanation to be correct. Put simply, yet another opportunity to provide scientific confirmation for a world beyond matter has turned up negative while the more parsimonious material model remains consistent with all the data.

We have already seen how the credibility of scriptures, religious leaders, and psychics would have been greatly enhanced if they made any risky predictions that came true or provided something that a disinterested party could find comparably convincing. Nobody has yet made such a case; neither has any one of the millions who have claimed to peer beyond the material world during prayer or meditation, or during an NDE, OBE, or other mystical experience whether induced naturally or artificially.

Some authors disagree.[66] They write about people who return from mystical journeys with information that they could not possibly have acquired by normal means. Surgeons will relate anecdotal tales of patients under anesthetic who later report details of the procedures performed on them. These may be explained by the fact that the senses, especially hearing, are not completely shut down by anesthetics. In any case, surgery is not the kind of suitably controlled experiment that can rule out such mundane explanations.

In 1968 parapsychologist Charles Tart reported that a young woman having an OBE in a "controlled" experiment was able to read a five-digit number placed above her head as she floated above her body.[67] However, it was later revealed that she could have read the numbers reflected in a wall clock.[68] Whether or not the subject actually used the clock, the possibility existed. This fact alone informs us that the experiment was not adequately controlled and should be disregarded. Like many of the parapsychologists and other paranormal researchers we have met, Tart failed to impose scientifically controlled conditions and then jumped to the conclusion he wished to make.[69]

Larry Dossey, whose questionable conclusions on the efficacy of prayer were discussed in chapter 9, has also written about NDEs in his book *Recovering the Soul*. There he tells the story of a women named Sarah who almost died when her heart stopped during a gallstone operation and was successfully resuscitated. He reports that she saw:

> a clear, detailed memory of the frantic conversation of the surgeons and nurses during her cardiac arrest; the OR [Operating Room] layout; the scribbles on the surgery scheduling board on the hall outside; the color of the sheets covering the operating table; the hairstyle of the head scrub nurse; the names of the surgeons on the doctors' lounge down the corridor who were waiting for her case to be concluded; and even the trivial fact that the anesthesiologist that day was wearing unmatched socks. All this she knew even though she had been fully anesthetized and unconscious during the surgery and the cardiac arrest.[70]

Dossey then produces the seeming clincher:

> But what made Sarah's vision even more momentous was the fact that, since birth, *she had been blind*. [Emphasis added]

Psychologist Susan Blackmore wanted to know more and contacted Dossey in March 1991. He wrote back:

> "Sarah's" story was a composite—the only composite story in the entire book, *Recovering the Soul*. My reasons for composing her were to dramatically illustrate the key features of nonlocal ways of knowing—ways that seem (to me) fully documented in the experiences of diverse numbers of human beings.[71]

Dossey thus admits that he made it all up! Do you still wonder why honest scientists have such contempt for so much of what is passed off as science in support of supernatural beliefs?

Other cases of reported NDEs by blind people have been tracked down by Ring. None have proved to be more than anecdotal tales. In 1987 he told Blackmore that "as much as this is the lore of NDEs, there has never, to my knowledge, been a case of a blind NDEr reported in the literature where there was clear-cut or documented evidence of accurate visual perception during an alleged OBE. (And you can quote me)."[72]

Some believers have been led to think that evidence for life after death exists in the studies of NDEs in cardiac arrest patients, such as those done by Sam Parnia and his associates in Southampton, England.[73] Even the CBS News, quoting a Reuters dispatch on June 28, 2001, reported, "A British scientist studying heart attack patients says he is finding evidence that suggests that consciousness may continue after the brain has stopped functioning and a patient is clinically dead."[74] However, when one reads the published article, there is no reference to the patients being "clinically dead," with irreversible loss of circulatory and respiratory functions or irreversible loss of neurological functions or the capacity for consciousness. Obviously such losses of functions were not irreversible because the patients were successfully resuscitated. Cardiopulmonary resuscitation is conventionally attempted in most cases of cardiac arrest. So it is very likely that the Southampton patients were all still "biologically alive," with some *cerebral* activity, and were not "brain dead" as it is conventionally defined.[75] In other words, they did not return from the dead.

Furthermore, the report was based on a total of sixty-three patients of which only seven reported memories and only four had what is clinically defined as an NDE according to a standard called the Greyson scale. None reported an OBE. This is hardly a sufficient sample from which to draw any significant conclusions. Indeed, if you read the paper, the authors' official conclusions are weak and unremarkable: "Memories are rare after resuscitation from cardiac arrest. The majority of those that are reported have features of NDE and are pleasant. The occurrence of NDE during cardiac arrest raises questions about the possible relationship between mind and brain. Further large-scale studies are needed to understand the aetiology and true significance of NDE."[76]

In a BBC News report on the Southampton research, Chris Freeman, consultant psychiatrist and psychotherapist at Royal

Edinburgh Hospital, said there was no proof that the experiences reported by the patients actually occurred when the brain was shut down. "We know that memories are extremely fallible. We are quite good at knowing that something happened, but we are very poor at knowing when it happened." It is quite possible that these experiences happened during the recovery, or just before the cardiac arrest. To say that they happened when the brain was shut down, I think there is little evidence for that at all."[77]

The story of the search for scientific evidence for a spirit world has been a sad one, filled with the sort of errors and falsities that can only give science a bad name. Susan Blackmore, who has managed to maintain her good name, recently wrote about her own thirty-year enchantment with psi:

> It was just over thirty years ago that I had the dramatic out-of-body experience that convinced me of the reality of psychic phenomena and launched me on a crusade to show those closed-minded scientists that consciousness could reach beyond the body and that death was not the end. Just a few years of careful experiments changed all that. I found no psychic phenomena—only wishful thinking, self-deception, experimental error and, occasionally, fraud. I became a sceptic. . . .
>
> So why didn't I give up then? There are lots of bad reasons. Admitting you are wrong is always hard, even though it's a skill every scientist needs to learn. . . .
>
> Another "psychic" turns up. I must devise more experiments, take these claims seriously. They fail—again. A man explains to me how alien abductors implanted something in his mouth. Tests show it's just a filling, but it might have been. . . . No, I don't have to think that way. And when the psychics and clairvoyants and New Agers shout at me, as they do: "The trouble with all you scientists is you don't have an open mind," I won't be upset. I won't argue. I won't rush off and perform yet more experiments just in case. I'll simply smile sweetly and say: "I don't do that anymore."[78]

NOTES

1. B. A. Brennen, *Hands of Light: A Guide to Healing through the Human Energy Field* (New York: Bantam New Age Books, 1988).

2. L. Richard Wheeler, *Vitalism: Its History and Validity* (1930; reprint, London: Witherby, 1930).

3. H. Driesch, *History and Theory of Vitalism* (New York: Macmillan, 1914); Henri Bergson, *Creative Evolution* (New York: Macmillan, 1911).

4. Peter Huston, "China, Chi, and Chicanary [*sic*]: Examining Traditional Chinese Medicine and Chi Theory," *Skeptical Inquirer* 19, no. 5 (1995): 38–42, 58.

5. Leonard Finegold, "Magnet Therapy," *Scientific Review of Alternative Medicine* 3, no. 1 (1999): 26–33.

6. Thomas S. Ball and Dean D. Alexander, "Catching Up with Eighteenth-Century Science in the Evaluation of Therapeutic Touch," *Skeptical Inquirer* 22, no. 4 (1998): 31–34.

7. For my detailed critique of paranormal claims, see *Physics and Psychics: The Search For a World Beyond the Sciences* (Amherst, N.Y.: Prometheus Books, 1990). Many references to other studies can be found therein.

8. For critical articles on alternative therapies, see *The Scientific Review of Alternative Medicine* (Amherst, N.Y.)

9. Because of a very bad law, most herbal remedies are not required to be tested by the U.S. Federal Drug Agency but are treated as "food supplements." Only recently have studies begun on the possible dangers of these substances, and many are being found to be both useless and dangerous.

10. Linda A. Rosa, "Therapeutic Touch," *Skeptic* 3, no. 1 (1994): 40–49; Bela Scheiber, "Therapeutic Touch: Evaluating the 'Growing Body of Evidence' Claim," *Scientific Review of Alternative Medicine* 1, no. 1 (1997): 13–15; Bela Scheiber and Carla Selby, eds., *Therapeutic Touch* (Amherst, N.Y.: Prometheus Books, 2000); George Ulett, "Therapeutic Touch: Tracing Back to Mesmer," *Scientific Review of Alternative Medicine* 1, no. 1 (1997): 16–18.

11. Martha Rogers, *The Theoretical Basis for Nursing* (Philadelphia: F. A. Davies, 1970); "Science of Unitary Human Beings," in *Explorations of Martha Rogers' Science of Unitary Human Beings*, ed. V. M. Malinski (Norwark: Appleton-Century-Crofts, 1986); Rogers, "Nursing: A Science of Unitary Human Beings," in *Conceptual Models for Nursing Practice*, 3d ed.,

ed. J. P. Riehl-Sisca (Norwark: Appleton & Lange, 1989); Rogers, "Nursing: Science of Unitary, Irreducible, Human Beings: Update 1990," in *Visions of Rogers' Science-Based Nursing*, ed. E. A. M. Barrett (New York: National League for Nursing, 1990).

12. Martha Rogers, "Nursing Science and the Space Age," *Nursing Science Quarterly* 5, no. 1 (1992): 27–34.

13. Joanne Stefanatos, "Introduction to Bioenergetic Medicine," in *Complementary and Alternative Veterinary Medicine: Principles and Practice* ed. Allen M. Schoen and Susan G. Wynn (St. Louis: Mosby-Year Book, 1998), p. 270.

14. Elissa Patterson, "The Philosophy and Physics of Holistic Health Care: Spiritual Healing as a Workable Interpretation," *Journal of Advanced Nursing* 27 (1998): 287–93.

15. Robert W. Loftin, "Auras: Searching for the Light," *Skeptical Inquirer* 14, no. 4 (1990): 403–409.

16. The use of the term "blackbody" to describe the Sun is confusing, but this is the technical designation for objects whose radiation primarily results from the thermal motion of its constituent particles. The cosmic microwave background is another "blackbody" at 2.7 degrees Kelvin.

17. Stefanatos, "Introduction to Bioenergetic Medicine," p. 228.

18. Ibid., p. 229.

19. Patterson, "The Philosophy and Physics of Holistic Health Care," p. 291.

20. For a sketch of the device and further discussion, see *Physics and Psychics*, pp. 237–41.

21. S. Ostrander and L. Schroeder, *Psychic Discoveries beyond the Iron Curtain* (Englewood Cliffs, N.J.: Prentice-Hall, 1970); Thelma Moss, *The Probability of the Impossible* (Los Angeles: Tarcher, 1974).

22. Ostrander and Schroeder, *Psychic Discoveries beyond the Iron Curtain*, p. 200.

23. John O. Pehek, Hay J. Kyler, and David L. Faust, "Image Modulation in Corona Discharge Photography," *Science* 194 (1976): 263–70; Barry Singer, "Kirlian Photography," in *Science and the Paranormal*, ed. George O. Abell and Barry Singer (New York: Scribners, 1981); Arleen J. Watkins and William S. Bickel, "A Study of the Kirlian Effect," *Skeptical Inquirer* 10, no. 3 (1986): 244–57; Watkins and Bickel, "The Kirlian Technique: Controlling the Wild Cards," *Skeptical Inquirer* 13, no. 2 (1989): 172–84.

24. Fritjof Capra, *The Tao of Physics* (Boulder: Shambhala, 1975); Gary

Zukav, *The Dancing Wu Li Masters: An Overview of the New Physics* (New York: Morrow, 1979); Robert G. Jahn and Brenda J. Dunne, "On the Quantum Mechanics of Consciousness, with Application to Anomalous Phenomena," *Foundations of Physics* 16 (1986): 721–72; Jahn and Dunne, *Margins of Reality: The Role of Consciousness in the Physical World* (New York: Harcourt Brace Jovanovich, 1987); Amit Goswami, *The Self-Aware Universe: How Consciousness Creates the Material World* (New York: G. P. Putnam's Sons, 1993).

25. Deepak Chopra, *Quantum Healing: Exploring the Frontiers of Mind/Body Medicine* (New York: Bantam, 1989); Chopra, *Ageless Body, Timeless Mind: The Quantum Alternative to Growing Old* (New York: Random House, 1993).

26. See my *The Unconscious Quantum: Metaphysics in Modern Physics and Cosmology* (Amherst, N.Y.: Prometheus Books, 1995).

27. Stefanatos, "Introduction to Bioenergetic Medicine," p. 228.

28. Ibid., p. 227.

29. Paul Kurtz, *The Transcendental Temptation: A Critique of Science and the Paranormal* (Amherst, N.Y.: Prometheus Books, 1986).

30. For more details and references on the material in this section, see Kurtz, *The Transcendental Temptation* and my *Physics and Psychics.*

31. C. E. M. Hansel, *ESP and Parapsychology: A Critical Evaluation* (Amherst, N.Y.: Prometheus Books, 1980); Hansel, *The Search for Psychic Power: ESP and Parapsychology Revisited* (Amherst N.Y.: Prometheus Books, 1989).

32. R. G. Jahn, B. J. Dunne, R. D. Nelson, Y. H. Dobbins, and G. J. Bradish, "Correlations of Random Binary Sequences with Pre-Selected Operator Intention: A Review of a Twelve-year Program," *Journal of Scientific Exploration* 11, no. 3 (1997): 345–67. Most of the experimental results and analyses can be found in the *Journal for Scientific Exploration,* a non-mainstream publication of the Society for Scientific Exploration whose purpose is to provide a forum for reporting unconventional and extraordinary studies. Details of earlier results, and a proposed quantum theory to explain the observations, can be found in Jahn and Dunne, *Margins of Reality.*

33. Jahn and Dunne, "On the Quantum Mechanics of Consciousness," and *Margins of Reality.*

34. J. E. Alcock, "A Comprehensive Review of Major Empirical Studies in Parapsychology Involving Random Event Generators or Remote Viewing," in *Enhancing Human Preformance: Issues, Theories, and*

Techniques (Washington, D.C.: National Academy Press, 1988), pp. 99–102; W. H. Jefferys, "Bayesian Analysis of Random Event Generator Data," *Journal of Scientific Exploration* 4 (1991): 153–69; Jefferys, erratum in *Journal of Scientific Exploration* 8 (1994): 255–56; Jefferys, "Response to Dobyns," *Journal of Scientific Exploration* 6 (1992): 47–57; Jefferys, "On p-values and Chance," *Journal of Scientific Exploration* 9 (1995): 121–22.

35. Stanley Jeffers, "Physics and Claims for Anomalous Effects Related to Consciousness," submitted to the *Journal for Consciousness Studies* (2001).

36. Helmut Schmidt, "Quantum Processes Predicted?" *New Scientist* (October 16, 1969): 114–15; Schmidt, "Non-causality as the Earmark of Psi," *Journal of Scientific Exploration* 7, no. 2 (1993): 125–32.

37. Morris Freedman, Stanley Jeffers, and Karen Saeger, "Intentionality, Random Processes, and Methodology," submitted to the *Journal for Consciousness Studies* (2001).

38. Jeffers, "Physics and Claims for Anomalous Effects Related to Consciousness."

39. Dean Radin, *The Conscious Universe: The Scientific Truth of Psychic Phenomena* (New York: HarperCollins, 1997).

40. I. J. Good, "Where Has the Billion Trillion Gone?" *Nature* 389, no. 6653 (1997): 806–807.

41. Douglas M. Stokes, *The Nature of Mind: Parapsychology and the Role of Consciousness in the Physical World* (Jefferson, N.C., and London: McFarland, 1997).

42. Douglas M. Stokes, "The Shrinking Filedrawer: On the Validity of Statistical Meta-Analysis in Parapsychology" *Skeptical Inquirer* 25, no. 3 (2001): 22–25.

43. A. Vaughn and J. Houck, "Intuition-Training Software: A Second Training Study," *Journal of the Society for Psychical Research* 64 (2000): 177–84.

44. Charles Honorton, "Meta-analysis of Psi Ganzfeld Research: A Response to Hyman," *Journal of the Society for Psychical Research* 64 (1985): 51–91.

45. Jeffrey D. Scargle, "Publication Bias ('The File-drawer Problem') in Scientific Inference" [online], http://arxiv.org/pdf/physics/9909033.

46. Kurtz, *The Transcendental Temptation*.

47. Raymond Moody, *Life After Life* (Covinda, Ga.: Mockingbird, 1975).

48. Susan Blackmore, "Near-Death Experiences," *Skeptical Inquirer*

16, no. 1 (1991): 34–35; Blackmore, *Dying to Live: Near-Death Experiences* (Amherst, N.Y.: Prometheus Books, 1993).

49. Kenneth Ring, *Life at Death* (New York: Coward, McCann & Geoghegan, 1980); Ring, *Heading Toward Omega* (New York: Morrow, 1986).

50. J. Morse, P. Castillo, D. Venecia, J. Milstein, and D. C. Taylor, "Childhood Near-Death Experiences: A Neurophysiological Explanatory Model," *Journal of Near-Death Studies* 8 (1986): 45–53.

51. Blackmore, "Near-Death Experiences."

52. Blackmore, *Dying to Live.*

53. M. A. Persinger, S. G. Tiller, and S. A. Koren, *Perceptual and Motor Skills* 90, no. 2 (2000): 659–74.

54. James Austin, *Zen and the Brain: Toward an Understanding of Meditation and Consciousness* (Cambridge: MIT Press, 1998).

55. Andrew Newberg and Eugene d'Aquili, *Why God Won't Go Away* (New York: Ballantine Books, 2001), p. 7; Sharon Begley, "God and the Brain: How We're Wired for Spirituality," *Newsweek*, May 7, 2000, pp. 50–57.

56. Newberg, *Why God Won't Go Away*, p. 2.

57. Ibid., p. 7.

58. Ibid., p. 5.

59. Ibid., p. 6.

60. Ibid., p. 129.

61. Ibid.

62. Ibid., p. 37.

63. Ibid., p. 9.

64. Ibid., p. 133.

65. Ibid., pp. 140–41.

66. Patrick Glynn, *God: The Evidence* (Rocklin, Calif.: Prima Publishing, 1997), pp. 111, 115.

67. Charles T. Tart, "A Psychophysiological Study of Out-of-the-Body Experiences in a Selected Subject," *Journal of the American Society for Psychical Research* 62 (1968): 3–27.

68. Leonard Zusne and Warren H. Jones, *Anomalistic Psychology: A Study of Extraordinary Phenomena of Behavior and Experience* (Hillsdale, N.J.: Lawrence Erlbaum Associates, 1982).

69. I give another example of Tart's poor experimental technique in *Physics and Psychics*, pp. 178–79.

70. Larry Dossey, *Recovering the Soul: A Scientific and Spiritual Search* (New York: Bantam, 1989), p. 18.

71. Blackmore, *Dying to Live*, pp. 131–32.

72. Ibid., p. 133.

73. Sam Parnia, D. G. Waller, R. Yeates, and P. Fenwick, "A Qualitative and Quantitative Study of the Incidence, Features, and Aetiology of Near Death Experiences in Cardiac Arrest Survivors," *Resuscitation* 48, no. 11 (2001): 149–56.

74. CBS News, "Mind over Matter," report of June 28, 2001.

75. Mihai D. Dimancescu, "Brain Death," *Newsletter of the International Coma Recovery Institute*, October/November 1984. The most recent version, June 1, 2000, is maintained on the Web site of the organization http://www.comarecovery.org/braindeath.htm.

76. Parnia, "Near Death Experiences in Cardiac Arrest Survivors."

77. BBC, "Evidence of 'Life after Death,'" report of October 23, 2000, GMT 10:24.

78. Susan Blackmore, "Giving Up the Ghosts: End of a Personal Quest," *Skeptical Inquirer* 25, no. 2 (2001): 25. A shorter version was published in *New Scientist*, November 4, 2000.

ELEVEN

THE PREMISE KEEPERS

The web of this world is woven of Necessity and Chance. Woe to him who has accustomed himself from his youth up to find something necessary in what is capricious, and who would ascribe something like reason to Chance and make a religion of surrendering to it.
—Johann Wolfgang von Goethe, 1795

FAITH IN SCIENCE AND GOD

A huge body of literature now exists in which authors with strong theological and scientific credentials argue that traditional religion, particularly Christianity, can be made consistent with all scientific knowledge. This new breed of scientist-theologians seeks to retain the fundamental Christian premise of a personal, loving, participating creator within a scientific framework. I will refer to them as the *premise keepers*.[1]

Rather than viewing science, or at least naturalism, as a mortal enemy to be defeated by all possible means, fair or foul, the premise keepers follow the lead of Newton and look to science to provide insight into the nature of God. Their efforts are far more credible than those who deliberately misinterpret or incompetently apply science to promote theological doctrines that they have formulated independent of any consideration of science.

The old creationists take the Bible literally and are forced to conclude that much of science is wrong. The new creationists claim that science is incomplete and requires an external "intelligent" designer to explain the complexity of the universe. By contrast, the premise keepers concede that established science, built upon a framework of materialism and naturalism, is empirically and theoretically sound.

The premise keepers embrace biological evolution as basically correct, recognizing that it is still evolving and that what disputes may exist among evolutionary scientists pose no serious challenge to the overall scheme. While a few premise keepers still trot out the fine-tuning cosmological argument for design in the universe as a whole, most are willing to admit that modern cosmology provides a viable model for a nonmiraculous origin of the universe. And while they disagree on many details, more theological than scientific, they generally view God as not so much interfering with natural processes, including chance, as working within them. As particle physicist, Anglican priest, and 2002 Templeton Prize–winner John Polkinghorne puts it, God does not work against the laws of nature because "that would be for God to act against God."[2] Still, one can only wonder how much of Christianity remains after its miracles are extracted.

The God of the premise keepers is not the Enlightenment deist god who set things in place at the creation and has since left us alone to live out his perfectly conceived divine plan. God still asserts creative control in the present world, but premise keepers dispute how much control and the mechanism God uses in interacting with the world. The premise keeper God is also not the pantheist god—some abstract Platonic concept of perfection and order. Polkinghorne rejects Spinoza's axiom *Deus, sive Natura* (God = nature), saying, "That was Einstein's God, but it is certainly not mine."[3] The God of the premise keepers is the God of the Bible, reinterpreted to assert his will within the natural laws, randomness, and chaos he purposefully wrote into creation. Whether that fits the God worshipped by most Jews, Christians, and Muslims is another matter.

Polkinghorne rightly points out that many physical notions, such as conservation laws, have been asked to carry too much metaphysical freight. He knows from his physics that conservation laws are simple symmetry principles. As Polkinghorne puts it, they are "consequences of the symmetries of creation and can easily be understood as expressions of the Creator's will rather than impositions on it." Alternatively, as I showed in chapter 8, these laws can just as easily be understood as the simplest assumptions that one can make, those that follow from the properties of the void. Polkinghorne also finds the vernacular use of the term *field* as a metaphor for omniscient spirit "quaint" and "distinctly limited."[4] We saw in chapter 10 that electromagnetic and quantum fields have been misappropriated as sources of a "human energy field" that some equate to the soul.

Polkinghorne also has little use for cosmological creation and fine-tuning arguments. He insists that "theology is concerned with ... ontological questions ... and gains little from science's fascinating, but largely theologically irrelevant, talk of temporal origins."[5] This places him outside the large group of theistic scientists, such as Hugh Ross and John Leslie, who are obsessed with cosmology and what they see as evidence that the universe has been fine-tuned for life.

Polkinghorne claims he is not engaged in "an apologetic exercise, trying to make the faith appear acceptable in a scientific age."[6] Following the tradition of Newton, he views science as a means to supplement scripture and revelation in learning about God.

What then remains as a basis to believe in this God? Polkinghorne can only fall back on personal tastes and desires. He sees human experience as encouraging belief in a divine mind, and in divine purpose behind the history of the world. He laments: "If cosmic history is no more than the temporary flourishing of remarkable fruitfulness followed by its subsequent decay and disappearance, then I think Macbeth was right and it is indeed a tale told by an idiot."[7] Perhaps it is just such a tale, whether Polkinghorne likes it or not.

CHAOS THEOLOGY

As we have seen, modern science has left theology in a quandary. The universe revealed by science shows humanity as an infinitesimal speck, with random chance as an important factor affecting events. Natural laws do not fully determine events but simply place constraints on them. Furthermore, these laws, such as the conservation principles, follow naturally from the symmetries of the void (chap. 8) and suggest the absence of purpose. Where can God exert his influence in such a universe?

As far as we know from current science, the development of macroscopic complex systems, most notably the structures of living organisms, arises by processes of self-organization and natural selection that include a large element of chance. Our present best guess is that the behaviors of macroscopic systems are *emergent phenomena* resulting from a blend of chance and constraint. Start them up again and they will not develop the same way because of the role of chance in randomly selecting out of many possible paths the particular path a system will follow as it develops with time. This includes, but is not limited to, the evolution of life and humanity, making it quite a theological challenge to find any purpose in it all without restoring determinism, natural or divine.

As we have noted in earlier chapters, many theists see the very existence of chance as a grave threat to their faith. The premise keepers take the opposite view; they look to chance as the place where God asserts his will. Polkinghorne and 2001 Templeton Prize–winner, biochemist, and fellow Anglican priest Arthur Peacocke have urged that interpretation. They think they have found room for God to act within the framework of *chaos theory*.[8] The two differ substantially in their specific theologies, their scientific perspectives arise from different backgrounds, and they do not view the role of chaos in exactly the same manner. Nevertheless, since they both weave their theologies with the common thread of chaos, I will group them together for this discussion.

The defining characteristic of chaotic systems is their extreme sensitivity to initial conditions, which results in their appearing to behave unpredictably. This is known as the "butterfly effect" in which, metaphorically, the flap of a butterfly's wings might affect the weather a week or so in the future. The butterfly effect was first discovered by meteorologist Edward Lorenz in 1960. He had written a computer program to model the atmosphere, but, given that computers were still in their infancy at the time, his model was necessarily crude. Still, Lorenz's model possessed feedback loops that were too complex to be analyzed by hand and, with the computer now available as a tool, this approach could be followed. His program produced outputs that were fed back as inputs, a notorious source of nonlinearity and instability.

Lorenz was surprised one day when his program gave different results after he had run it over again from what he thought was the same starting point. He soon realized that the computer had taken the number 0.506127 that was stored in its memory and rounded it off to 0.506 in a printout. Lorenz had then reentered it as 0.506000, and the results came out grossly different.[9] Thus, this seemingly insignificant difference of 0.000127 in a single input variable dramatically changed the nature of his overall results in modeling weather.

Since then, chaos theory and its more aptly named cousin, *complexity theory*, have provided important new insights into the nature of classical (nonquantum) physical systems, including many that surround us in everyday life. In general, a system of three or more bodies can exhibit chaotic behavior if it is *nonlinear*.

That means, for example, if you drive the system with some input and double that input, the effect will not necessarily be doubled. If you push a child's swing with a certain force, and then double that force, the amplitude of the swing will double when the initial push is very gentle and the amplitude is small. But keep doubling the force and soon the response will no longer be linear. Making sure no one you like is on the swing, you can even push it over the top. The simple swing, or, more technically, the damped, driven, non-linear pendulum, exhibits all the basic features of a chaotic system.

Besides the butterfly effect, chaotic systems exhibit other interesting properties. Important for our purposes, they can remain for some time in a quasi-stable state; but, then, a small perturbation can drive them to a completely different state. The atmosphere is again a good example. While the flap of a butterfly's wings probably won't do it, some small air movement or abrupt temperature change in one place can drive the weather over a large area from a quasi-stable calm state to a quasi-stable stormy state.

Evolution may also exhibit elements of chaos, as a small change in the environment results in the relatively abrupt movement of an ecosystem from one state to another, perhaps explaining the *punctuated equilibrium* that some paleontologists, such as Stephen Jay Gould and Niles Eldredge, claim occurs in evolution.[10] Many biological and even social systems exhibit the properties of chaos. One of the most studied mathematical examples is the *logistic map*, which is a very simple model of population growth limited by resources.[11]

Polkinghorne and Peacocke see chaotic systems as providing an opening for God to act in the world without having to violate any natural laws, or at least not violating them in any noticeable way (and thus confirming God's existence). Neither visualizes God as selectively injecting huge amounts of *energy* into various places in the universe needing his intervention, thus violently breaking the law of conservation of energy. Rather, in Polkinghorne's scheme, he injects *information*. God provides a gentle nudge that moves a complex system along the path he wishes it to go, taking advantage of the amplifying effect of chaos.

Although this type of divine intervention by information injection does not involve significant energy transfer, it must involve some. This would still violate energy conservation. Furthermore, as we saw in chapter 4, information is equivalent to entropy, or at least changes in entropy. Thus, God's intervention by the injection of information would amount to a violation of the second law of thermodynamics, with all the entropy of a process not accounted for within our universe. Like the flap of a butterfly's wings, however, the divine input could be so small as to be unmeasurable. Chaos, in Polkinghorne's scheme, thus provides a means for God to act in an undetectable way; but his act would still violate natural law.

Peacocke's vision of the role of chaos differs from Polkinghorne's, although Polkinghorne has not explicitly rejected the former's approach. Peacocke does not imagine God interfering in any specific event but acting on the whole by a process called *top-down causality*. In a trivial example, if you rotate a wheel, you are causing all the atoms of the wheel to move in a circle.

Thomas Tracy notes that Peacocke's proposal hinges on "the supposition that top-down explanations cannot be analyzed in terms of structures of bottom-up explanation."[12] Tracy points out that Peacocke makes "the move from the whole-part explanation to treating the whole (or the nature of the system) as a cause." In other words, Peacocke is asserting some new holistic principle.

Such a principle had been proposed earlier by Nobel laureate chemist Ilya Prigogine and his collaborators, with no theological intent.[13] For example, consider the arrow of time discussed in chapter 7, which we saw emerges as a principle for many-body systems on the macroscopic scale. Prigogine agrees with this, but further conjectures that it acts down to the elementary particle level, giving time a direction on that scale.[14] However, as I explained in *Timeless Reality*, all elementary processes allow for time reversibility and, indeed, quantum phenomena exhibit backward-time causality.[15] If Prigogine were correct, we would expect to see time irreversibility at the quantum scale. In fact, we do not. Other conjectures about top-down causality have similarly failed to yield empirical support.[16] So top-down causality seems to be a poor metaphor for God's action in the world, if that action is to be seen as in harmony with natural processes.

Peacocke does not base his case on Prigogine's model. The Templeton laureate simply mentions that Prigogine and others have demonstrated that "it is the interplay of chance and law which is in fact creative with time, for it is the combination of the two which allows new forms to evolve."[17] Of course, there is nothing new in this. What is new is the notion that some principles exist that cannot be understood in the traditional scientific reduction of a system to it parts.

In fact, bottom-up explanations for all physical systems, including chaotic ones (and the rotating wheel), are conventionally made. Reductionism—in which you analyze a system in terms of its parts—does, in fact, recognize that parts can interact, creating unique systems containing novel and unpredicted properties. Reductionist analyses do lead to just this outcome. In particular, behavior of chaotic systems was discovered by completely reductionist methods in which the parts of a system and their mutual interactions were modeled on a computer. The computer programs that produced these results used nothing but reductionist physics principles. No new fundamental physical principles, reductionistic or holistic, were either hypothesized or uncovered in the process. Far from destroying materialist reductionism, as some science writers have breathlessly reported,[18] chaos and complexity theory have provided further validation for the model in which all phenomena arise from elementary particles and their interactions. The only place where some dispute remains among scientists is in the workings of the brain and how it produces what we know as consciousness. This is not to say that everything is predictable from elementary particle theory. As already noted, chance plays such an important role in all physical phenomena that such predictability is not to be expected.

For Polkinghorne, "The most significant event in cosmic history to date [was] the dawn of consciousness."[19] Here again, he calls on chaos and complexity theory to provide an explanation. He asserts that "holistic and relational concepts are coming to play an increasing role in science." These are regarded as "congenial to theological thinking," as exemplified by "much Trinitarian discussion that emphasizes relationship (communion) as the ground of being."[20]

Polkinghorne argues that "mathematics is the exploration of an existing noetic realm."[21] This agrees with the line prompted by his fellow Oxfordian, mathematician Roger Penrose, that "there is something absolute and 'God-given'" about mathematical truth.[22] Admitting the influence of his personal yearnings, Polkinghorne says: "I believe there is a much more persuasive case for believing in the reality of the Mandelbrot set than in the reality of the Idea of a lion." (The Mandelbrot set is a mathematical contrivance with fascinating properties, including chaos, which is discussed in detail by Penrose and in many books on chaos.) According to Polkinghorne, the realm of physical and mental experience are "parts of an interlinked complementary created reality."[23] This may not be traditional Platonism, but it is Platonism nonetheless. Penrose, however, explicitly disassociated his ideas on consciousness from religion.[24]

THE NEW SCHISM

Although I am trying to avoid theological squabbles and stick to what I know best, the science, I must at least mention the sharp theological differences among the premise keepers, and even greater disputes between them and other Christian thinkers. These various internal schisms are more formidable than any that may separate them from scientists and bear little resemblance to those that have marked the history of the Church.

I mentioned that Polkinghorne and Peacocke differ substantially in their theologies. Polkinghorne holds on to rather conservative beliefs, such as the Virgin Birth and Resurrection, while Peacocke questions many traditional teachings. After Peacocke was awarded the million-dollar 2001 Templeton Prize for Progress in Religion, *Christianity Today* editor-at-large John Wilson commented:

Peacocke's distinguished career as a scientist included early research on DNA. His theology, alas, is typical of that of many thinkers who set out to explain what is "credible" for us to believe about God and his "interaction with the world" given the "comprehensive, indeed dazzling, perspective on the being and becoming of the world and of humanity that the sciences have . . . unveiled to our generation." This turns out to entail a rejection of anything resembling Christian orthodoxy from the first century to the present.[25]

Thus, from across the canyon that separates him from me, Wilson echoes the question that I have raised several times already: What is left of Christianity when it is pruned of virtually every traditional teaching?

QUANTUM THEOLOGY

One of the major confusions about chaos theory results from the fact that it is grossly misnamed. The term "chaos" implies randomness or lack of order. In fact, chaotic systems in physics are perfectly orderly. "Deterministic chaos" is more apt. It describes what appears when one takes the equations of classical mechanics that apply to nonisolated, many-body systems and uses a computer to solve them iteratively, that is, in a series of steps in time. Deterministic chaos also happens with purely mathematical objects such as the logistic map or Mandelbrot set. Since the paths followed by the system are very sensitive to the initial conditions, the results look "chaotic." You start the system in one place, and you get one path. You slightly change the starting point, and you get a completely different path. However, when you take care to always start the system in exactly the same place (do not round off your numbers the way Lorenz did), the system always follows exactly the same path and ends up in exactly the same place. In other words, mathematically, *chaotic systems are perfectly predictable*. Run the computer program once, and you can then predict the path the system will follow, with 100 percent certainty, each time you run the program again from the same, exact initial point.

Before the advent of quantum mechanics, the physical universe seemed to be a vast clockwork mechanism that operated according to the equations of Newtonian mechanics. In Newtonian mechanics, the motion of a particle is fully determined by its initial position, momentum, and the forces acting on it. Since everything in the material universe, including the human body, is made up of particles, then no room for choice or chance existed within this framework.

This is why so many educated people during the Enlightenment period that followed the development of Newtonian science became deists, although, as we have seen, Newton himself did not. The deists saw no place for a god to act in the present world once he had set it all in motion at the creation. Newton believed that the God of the Bible acted in the gaps where mechanics provided no explanation, although this notion was rejected by his most important successors, notably Laplace, who saw all the gaps as potentially being filled.

However, free will, both human and divine, comprises an essential aspect of the monotheistic religions. Unless people have free will, sin and virtue are meaningless. Unless God has free will, miracles cannot occur and prayers are not answered. Both human and divine free will were threatened by the deterministic world view of classical physics.

As quantum pioneer Werner Heisenberg himself realized, the uncertainty principle that he developed in the 1920s provided a possible way out of the Newtonian world machine. Heisenberg showed that since the position and momentum of a particle cannot be measured simultaneously with unlimited precision, particle motion is ultimately indeterminate.

Nancy Murphy and other theologians have raised objections to the use of classical chaos as a medium for God's action.[26] She notes, as I have above, that classical chaos is a deterministic theory and so still has no room for God's action. However, quantum mechanics *can* eliminate the predictability of deterministic chaos when one is dealing not with a computer simulation but with an actual physical system.

All physical systems are fundamentally quantum in nature, but for most macroscopic examples of common experience, the classical limit of quantum mechanics provides a good approximation, and these systems behave deterministically to a good approximation. What I will operationally define as *true randomness* in otherwise deterministic chaotic systems can come about when the Heisenberg uncertainty principle prevents the initial variables of the system from being measured with sufficient accuracy for the evolution of the system to be predictably repeated. In this way the motion of a particle that is part of a chaotic system becomes intrinsically unpredictable even though the system itself is deterministic.

Murphy also objects that Polkinghorne "has not provided a clear account of *how* God works on the inside, in the process." She then proceeds to give her own account of how God does it:

> I take it that . . . it is necessary that God work in the inside of *all* created entities—which must mean in turn that God works within the smallest constituents of macroscopic entities, since these smallest constituents are entities of their own right. If we begin with this hypothesis, it is not necessary—in fact it is counterproductive—to argue for causal indeterminism at higher levels of organization (excluding the human level) since God's will is assumed to be exercised by means of the macro-effects of subatomic manipulation.[27]

As I indicated in the previous section, as far as we know from current physics the structures of complex, many-body systems, such as biological organisms, are not fully determined by the laws that govern the behavior of elementary particles. Chance plays a role at all levels, from elementary particles to atoms to molecules to biological systems, and so on. But in order for that chance to be fully indeterministic, it must arise from quantum interactions. So, if I interpret Murphy correctly, she is saying that God would still have to exert his control at the microlevel. For example, God could have decided just when a potassium nucleus in the blood of some early mammal decayed and the beta electron from the decay knocked another electron from an atom in its DNA, thus producing one of the millions of mutations that eventually led to the evolution of *Homo sapiens*.

The other place where quantum mechanics opens up a gap for God's action is in the entanglement of observer and observed, mentioned in chapter 10. This is not discussed much in Christian literature, perhaps because of its association with New Age ideas and Eastern mysticism.[28] Quantum mysticism is based on the notion that consciousness can affect reality by controlling the collapse of the quantum *wave function* which determines the state of a physical system.

In conventional applications of quantum theory, the wave function is a mathematical abstraction that allows one to calculate the probability that a system will be found in a particular volume at a particular time. The evolution of the wave function with time is governed by the famous Schrödinger equation, which I need not present here. Suppose the "system" is a single electron emitted by an atom and has an equal probability of going in any direction. The wave function will mathematically represent that fact. Now, further suppose that we have a detector at one end of the laboratory that covers only a fraction of the solid angle, say 1 percent. That is, on average, only one in a hundred electrons will trigger that detector. If that detector registers a hit, this tells us that the electron went in that particular direction. At the moment of detection, the probability that the electron will be found in the detector volume abruptly changes from 1 to 100 percent. The probability that the electron will be found elsewhere drops to zero. The wave function is then said to "collapse" to reflect the change of information we have about the system, from it-could-be-anywhere to it's-precisely-right-here.

Since humans build the measuring equipment and ultimately analyze the data, some have suggested that it is an act of human consciousness that collapses the wave function. Although the wave function, in the conventional interpretation of quantum mechanics, is just a mathematical quantity, it must bear some relationship to reality since it does enable us to calculate probabilities of things happening with exceptional precision. In a Platonic worldview, that wave function is more of an objective reality than the electron it describes, in which case consciousness, or at least the detection apparatus, acts to control reality!

As I showed in *The Unconscious Quantum* and *Timeless Reality*, several alternative, nonmystical interpretations of the admittedly strange behavior of quantum systems are available. Nevertheless, many authors claim that wave function collapse provides a mechanism by which human consciousness can connect, instantaneously, to a cosmic field of force pervading the universe. While Christian theologians might be tempted to associate this field with the spirit of God, the premise-keeper theologians I have cited here are much too scientifically knowledgeable to fall into the bottomless pit of quantum mysticism.

EVOLUTION THEOLOGY

A number of believing scientists and science-savvy theologians have woven Darwinism into their metaphysical schemes. They have taken the apparent role of chance in complex systems to be gaps into which they can insert their God. In this they differ from the chaos and quantum theologians, who still envisage God as interfering with the process—albeit in an undetectable way.

In chapter 2, I mentioned biologist-theist Kenneth Miller's devastating critique of both the new and old versions of creationism in *Finding Darwin's God*. He sees creationists as missing a very important feature of the randomness involved in evolution, which they so abhor:

> The only alternative to what they describe as randomness would be a nonrandom universe of clockwork mechanisms that would also rule out active intervention by a supreme Deity. Caught between these two alternatives, they fail to see that the one more consistent with their religious beliefs is actually the mainstream scientific view linking evolution with the quantum reality of the physical sciences.[29]

Of course, mere consistency with a specific religious belief is not a very powerful argument for that belief. Randomness is equally consistent with a religion based on the Tooth Fairy, or no religion at all.

As discussed above, chance, or indeterminism, in science rescues theology from the old deism. If the Newtonian world machine had not been dismantled by quantum mechanics, the only place that God need have acted in the world was at the creation. Miller agrees with physicist and 1999 Templeton Prize–winner Ian Barbour that "natural laws and chance may equally be instruments of God's intentions. There can be purpose without an exact predetermined plan."[30]

Two kinds of purpose can still be conceived in this context. In one, described above, God asserts his purpose by micromanaging the quantum events that induce the various choices on the path a system follows as it develops with time. In this case, things still come out exactly the way God wants, including the evolution of humanity. This still complements the traditional belief that we are special creations, formed in "God's image," if not exactly physically, then at least spiritually with his personal qualities of love and forgiveness.

The second possibility, which one hears from the evolution theologians, is that God does not micromanage but allows chance to operate. In this case, his purpose is served by any path that is followed, including, it seems, a path that does not lead to the evolution of humankind.

In *God After Darwin*, theologian John Haught writes:

A God whose very essence is to be the world's open future is not a planner or a designer but an infinitely liberating source of new possibilities and new life. It seems to me that neo-Darwinian biology can live and thrive quite comfortably within the horizon of such a vision of ultimate reality.[31]

In this theology, the accidental processes of nature are *the means* by which God allows freedom to exist in the universe. The future is in fact open and not predetermined by physics or God.

CHANCE DISPUTED

Christians, however, are far from unanimous on the role played by chance in the universe. Many find a God with no special place for humanity very difficult to swallow. In *A Case against Accident and Self-Organization*, Christian attorney Dean L. Overman attempts to defeat that prospect by disputing any place for chance in both the formation of the universe and the evolution of life. He asks: "Is it mathematically possible that accidental or chance processes caused (a) the formation of the first form of living matter from non-living matter and (b) the formation of a universe compossible with life?"[32]

Overman concludes that the probability is so small as to be impossible. Life would never have evolved if it was all left to chance. We have seen this type of argument several times before in this book and have noted that a low probability for any particular event does not make that event impossible. Overman asserts that "mathematicians normally regard anything with a probability of less than one in 10^{50} as mathematical impossibility."[33] This is simply incorrect. All one has to do is visit one of the Web sites that provide random numbers and write down the first 50 digits. In a minute or two one can produce a sequence that had prior probability of 1 in 10^{50} of being due to chance.

Recall from chapter 4 that William Dembski has attempted to make a similar argument with his design inference filters. He introduces the notion of complex specified information, where complexity is defined as a sequence with chance odds of 10^{150} to 1. While this is a far stronger criterion than that used by Overman, it is still insufficient for inferring design in and by itself. Dembski knows this and so adds the criterion of *specification*. The information sequence must contain some sort of "message" that could not have occurred by chance or other natural processes. When that message is specified *in advance* and it then appears, *then* a strong inference can be drawn that some intelligence or artifice was involved. For example, if we specified in advance that a message contain 150 specific digits, such as all 7s or a 150-digit repeated sequence such as 1234512345, ... then it would indeed be highly improbable to obtain this message from the first 150 random numbers drawn off the Web. Much more dubious, however, is the process of looking at some sequence after the fact and then declaring it to be *specified*.

In any case, neither Overman nor Dembski can infer intelligent design by simply calculating a low probability for some physical event being undesigned without comparing that probability with one calculated for the alternative hypothesis of design. How do they know that the design probability is higher? What is the probability that God was produced by chance? Given the clear absence of design in so many phenomena, and the lack of evidence for any external designer, one might reasonably conclude that it is, in fact, the probability of design that is the smaller.

Even theists have found Overman's case against accident and self-organization unconvincing. Writing in *Christianity Today*, Rebecca Flietstra calls Overman's book "a misguided quest for proof" that "contains some crucial misunderstandings of scientific method."[34] She points out that arguments that may work well for law do not work well at all in science. According to Flietstra, Overman uses "inference from the universal to the particular" and claims that "inference from the particular to the universal is not valid." She notes that "science is not law; what is necessary for the lawyer is impossible for the scientist. That is, every experiment is a study of the particular, and every theory an 'inference from the particular to the universal.'"

For example, Overman follows the lead of many creationists in dismissing the famous experiment done by Stanley Miller in 1953 in which amino acids were produced by an electrical spark in a gaseous mixture.[35] Since the mixture did not exactly duplicate what we now think was the atmospheric composition at that time, Overman argues that it is logically incorrect to conclude that life began in this way. Flietstra comments:

> But what Overman describes as a logical misstep is more accurately understood as a result of an incomplete understanding of abiogenesis. And so, while it certainly is a leap to move from

Miller's experiments to a full-fledged explanation of life's origins, that does not mean it is impossible for us to derive any information from his experiments.[36]

I would add that Miller's results were of immense significance, not because they demonstrated exactly how life came about but because they showed that amino acids, very complex molecules that are critical components of life on Earth, are easily fabricated from much simpler ingredients by purely natural processes. Indeed, radio astronomy has identified complex molecules in interstellar gas clouds, and hydrogen cyanide, formed by natural processes in the atmosphere, can naturally polymerize and form amino acids.

Fifty years before Overman, Dembski, and other modern creationists claimed to prove that it was statistically impossible, Miller's experiment had already provided a beautiful empirical example of natural processes creating precisely the kind of information inherent in living systems. They might as well try to prove that it is impossible for humans to travel to the moon.

Flietstra also criticizes Overman's probability calculations, which greatly distort the processes of evolution. For example, he computes the odds of producing a single bacterium by chance at one in $10^{40,000}$, completely ignoring the role of natural selection. Of course, evolutionary biologists do not claim that bacteria are produced by random molecular collisions, but by natural selection in concert with chance.

NATURAL THEOLOGY

A huge chasm exists between intelligent design "science" and evolution "theology." The basis of that disagreement is fundamental, but those fundamentals are theological rather than scientific. As we have seen, the scientific claims of the design theorists have little merit and are, in many cases, demonstrably wrong. One can only wonder why they are promoted with such fervor, and despite protestations to the contrary, the reason is clearly the promotion of traditional religion. An evolution theology in which God does not poke in his finger to help things along seems to allow the possibility that humanity might not have developed at all, that we are just an accident and God's plan could have been fulfilled in many other ways—perhaps with no sentient life whatsoever.

The premise keepers not only support evolution, they regard with distaste the attempts of new and old creationists alike to cast doubts about its validity in the public forum. In a speech before the American Academy of Religion, Barbour had this to say about intelligent design:

> Philosophical proponents of intelligent design, such as William Dembski and Stephen Meyer, write in the tradition of natural theology in which science is used as evidence of the existence of a designer. My own approach is not natural theology but a theology of nature in which one asks how nature as understood by science is related to the divine as understood from the religious experience of a historical community.[37]

Dembski hotly disputes this assessment, arguing that, in fact, the opposite is true—that it is Barbour and his proevolution cohorts who are the natural theologians.[38]

What is this "natural theology" that is to be avoided like the plague by modern theologians? Dembski informs us that it had its origin in the eight pre-Darwin Bridgewater treatises that were financed by the Rev. Francis Henry Egerton, eighth and last Earl of Bridgewater, who died in 1829. He willed that £8,000 be used by the president of the Royal Society of London to publish works on:

> ... the Power, Wisdom, and Goodness of God as manifested in the Creation illustrating such work by all reasonable arguments as, for instance, the variety and formation of God's creatures, in the animal, vegetable, and mineral kingdoms; the effect of digestion and thereby of conversion; the construction of the hand of man and an infinite variety of other arguments; as also by discoveries ancient and modern in arts, sciences, and the whole extent of modern literature.

Note the charge was to illustrate and advance an already established notion, divine creation, rather than open up questions that might lend doubt to that notion. This is reminiscent of the Templeton charge, quoted in the preface, to explore "human potential within its ultimate purpose."[39]

As Dembski elaborates:

> The stereotypical argument of a natural theologian begins with "Isn't it amazing how . . . ." The natural theologian then fills in the blank with some feature of the natural world that inspires admiration and argues how this feature, once properly interpreted, demonstrates the manifold wisdom, power, and goodness of God. The problem with such arguments, of course, is that they can be turned on their head. Thus for every instance where the natural theologian finds reason to sing God's praises, the natural anti-theologian finds reason to lament nature's cruelty. Darwin, for instance, thought there was "too much misery in the world" to find solace in natural theology: "I cannot persuade myself that a beneficent and omnipotent God would have designedly created the Ichneumonide with the express intention of their feeding within the living bodies of Caterpillars, or that a cat should play with mice." Other examples he pointed to included "ants making slaves" and "the young cuckoo ejecting its foster-brother."[40]

Dembski argues that the proevolution science-theists, such as Polkinghorne, Barbour, and Peacocke are "as much engaged in natural theology as any natural theologians of time past." For the only time in this book, Dembski and I find ourselves in agreement. Polkinghorne, for example, insists that he is not just engaged in apologetics but looking to science to gain further insight into a God that he has already assumed exists. The other premise keepers operate from the same basis.

Dembski tries to extract his own specialty, intelligent design, from the morass of natural theology:

> The fundamental idea that animates intelligent design is that events, objects, and structures in the world can exhibit features

that reliably signal the effects of intelligence. Disciplines as diverse as animal learning and behavior, forensics, archeology, cryptography, and the search for extraterrestrial intelligence thus all fall within intelligent design.[41]

In other words, Dembski claims that he and his colleagues in the intelligent design movement are not looking to understand the nature of God from the observations of science. Rather, they are seeking to show that science cannot proceed to explain all phenomena without recourse to external, intelligent design. The fact that they have so far failed and made several fundamental scientific errors is beside the point. Perhaps they have simply not yet found the smoking gun.

Dembski attempts to distance himself further from theology of any sort by stating:

> The designer of intelligent design is not the God of any particular religious faith and not the God of any particular philosophical reflection but merely a generic intelligent cause capable of originating certain features of the natural world. Positing such a designer to account for certain types of biological complexity is like positing quarks to account for certain properties of subatomic particles. The point is to see what a designer helps explain; the point is not to establish the existence of the designer.[42]

I have already noted, in chapter 2, that Dembski and his fellow new creationists have learned from the mistakes of the old creationists and are careful not to make overtly religious claims that will be tossed out in court. Their position here is clearly strategic. We have seen that the intelligent design movement is largely political, rather than scientific, and by admitting no religious purpose, Dembski and his colleagues expect they can install their brand of antievolutionism into school curricula. If successful, they would be insinuating religion into science through the back door. Hopefully, the courts won't be so easily deceived, although many politicians and school board officials have already taken the bait.

Dembski has an advanced degree in theology, and his books, along with those of most others in the intelligent design movement, are clearly motivated by his Christian beliefs. The subtitle of Dembski's most important book, *Intelligent Design*, reads *The Bridge between Science & Theology*. The intelligent design movement is supported by the collective financial resources of wealthy, committed Christians. No one involved leaves any doubt whom they regard as the intelligent designer: the God of the Bible. Their financial supporters are not putting up big money to disprove God's existence. And no one should be so naive as to believe design theorists when they say that their purpose is innocent of any religious motivations.

Ignoring his own writings on the subject, Dembski disingenuously blames proevolution Christians, such as Miller and Haught, for bringing theology into the discussion:

> The juxtaposition here of God and Darwin is not coincidental. I submit that the preoccupation by critics of intelligent design with theology results not from intelligent design being inherently theological. Instead, it results from critics having built their own theology (or antitheology as the case may be) on a foundation of Darwinism. Moreover, because intelligent design challenges that foundation, critics reflexively assume that intelligent design must be inherently theological and have a theological agenda. . . . Critics of intelligent design resort to a classic defense mechanism in which they project onto intelligent design the very thing that intelligent design is unmasking in their own views, namely, the extent to which Darwinism, especially as it has been taken up by today's intellectual elite, has itself become a project in theology.[43]

A few years ago, physicist-theist Howard Van Till found himself enmeshed in a similar debate with intelligent design mouthpiece Phillip Johnson, on the pages of the Christian journal *First Things*.[44] Van Till promotes an evolution theology akin to that which I described above, in which God does not micromanage:

> In the context of this traditional Christian vision of God's creative work (notably different from Johnson's theokinetic picture), we

might now wish to employ the vocabulary of twentieth-century science and speak about the full array of functionally viable forms of DNA (and the creatures thereby represented) as constituting a "possibility space" of potential life-forms—this possibility space itself, along with all connective pathways, being an integral component of the world brought into being at the beginning. Furthermore, in the language of this theistic paradigm of evolutionary creation we would speak of DNA being enabled by the Creator to employ random genetic variation as a means to explore and discover (in contrast to create) viable pathways and novel life-forms so that the Creator's intentions for the formative history of the Creation might be actualized in the course of time.

See, then, what this evolutionary creation paradigm accomplishes: Do material processes have to create? No, the possibility space of viable and historically achievable life-forms is an integral aspect of the world that God created at the beginning. Material systems need only employ their God-given functional capacities to discover some of the possibilities thoughtfully prepared for them. But, one might ask, how can such "mindless" material processes function to bring about what appears to be the product of "intelligent design"? The point is that they are not really mindless at all. Rather, every one of these processes and every connective pathway in the possibility space of viable creatures is itself a mindfully designed provision from a Creator possessing unfathomable intelligence.

But Johnson is not buying this. To him, this "theistic naturalism" is not Christianity. Rather it is science attempting to co-opt the revealed truth with nothing more than an updated deism:

I do think that theistic naturalism is ultimately incoherent, but the incoherence is not obvious and it is understandable that many rational and intelligent Christians have overlooked it. First, we have all been taught to think of "science" as a neutral, objective process of fact-finding that is not biased in favor of a comprehensive metaphysical naturalism. In consequence the conclusions of science must be accepted by anyone who wants to be considered rational by the standards of the academic world.

When "science says" that natural selection can accomplish wonders of creativity, that is the end of the matter. Religion cannot survive in a naturalistic academic culture if it opposes science, and so religion must accommodate to science on the best terms it can get. Effectively, that means that God must be exiled to that shadowy realm before the Big Bang, and He must promise to do nothing thereafter that might cause trouble between theists and the scientific naturalists.

In short, theistic naturalism is best understood as an intellectual strategy for coping with a desperate situation. It was barely tenable as a philosophical position as long as the leading scientists believed, or pretended to believe, that science is a limited research activity which does not aspire to occupy the entire realm of knowledge. Today many of the world's most famous physicists are proclaiming the imminent prospect of a "theory of everything"—and they do mean everything. It may be that these physicists—and the evolutionary biologists who talk just like them—are no longer practicing "science" and have become metaphysicians. What is important is that they mix metaphysics and science together and present the whole package to the public with all the awe-inspiring authority of science.

While Johnson obviously wants to pick a fight with atheistic scientists, it seems that the critical battle for him is not between religion and science but between different viewpoints within Christianity—between antiscience and proscience theologies. Perhaps this is why so few atheistic scientists have taken Johnson's bait; they see nothing of scientific merit to argue with him about and, being wiser than I, leave theology to the theologians.

The premise keepers deserve credit for working with established science instead of fruitlessly challenging it. Johnson and his cohorts will not succeed in their efforts to undermine evolution theory and naturalism, at least within the scientific community. They might as well try to undermine the atomic theory of matter.

Still, Genesis says that God made Man in his image to rule over the rest of creation. Dr. David A. Staff, minister of the First Evangelical Free Church in Ames, Iowa, expresses the widely held,

orthodox Christian view that Man is "God's very special creation
. . . the object of God's stunning, loving plan."[45] The premise
keepers who reject guided evolution, who accept that humanity is
an accident, have not yet succeeded in molding their God into the
traditional God of Christianity.

THE ALTAR OF HYPOTHESIS

Willem Drees is, like John Polkinghorne, another physicist-turned-
theologian. He also seeks to establish criteria by which traditional
ideas of God can be brought into consonance with science—in particular, modern physics and cosmology.[46]

Drees recognizes that science provides no evidence for the
existence of God and that nothing in current cosmology demands
that the universe was purposefully created—a refreshing admis-
sion from a theologian. Drees accepts that God is a postulate motivated more by human desires than scientific facts. He admits that
theology, not science, carries the burden here and hopes that by
"deeply immersing ourselves in science, rather than stopping
short of the innermost sanctuaries, the tradition may show up in a
new light."[47]

Drees does not attempt to exploit the limitations of science as
openings for theology to fill. He recognizes that science has a way
of eventually filling its own gaps. Besides, Drees does not regard a
God of the gaps as a desirable metaphor to describe his postulated
God. His God is transcendent, beyond space and time and not in
the business of filling every nook and cranny of the universe.

Drees finds the fine-tuning arguments for a universe designed
for humankind wanting on a number of fronts. (It is interesting
that, among the premise keepers, most of the physicists and cosmologists such as Polkinghorne and Drees have no truck with the
fine-tuning argument; on the other hand, biologists like Miller,
who know little cosmology, still think there is something to it.)
Drees points out that science may someday find the physical laws
that explain these coincidences. The fine-tuning God is again a

God of the gaps. Drees also views the ideas of Polkinghorne and Peacocke, that God acts via chaos theory, and other suggestions that God acts via the uncertainties in quantum mechanics, as equally futile attempts to seek the God of the gaps.[48]

Drees discusses the proposal made by physicist Frank Tipler in *The Physics of Immortality* that some form of intelligent life will be able to achieve eternal life.[49] Tipler imagines a single pocket of increasingly higher-level organization evolving to the ultimate *Omega Point* god, envisioned by Jesuit priest and paleontologist Pierre Teilhard de Chardin.[50] Tipler asserts the "Final Anthropic Principle," proposed with John Barrow,[51] which declares that intelligent information processing once brought into existence will never die out. Tipler suggests that the Omega Point will contain all of past life within its information-processing system, and thus every being who ever lived will become resurrected at that time—as a computer simulation. Not only that, everything that can be known will be known, and so not only will each of us live our lives over and over an infinite number of times in an instant, we will live all the other lives we might have lived, experience every conceivable pain and pleasure.[52]

If ever there was a materialistic theory of heaven and hell, handed on a platter to those who want to believe in God, this is it. But Drees does not buy this version of scientific theology either. He sees no point in the concept of a deity so far in the future that it can have no possible current relevance, especially since we will all be resurrected anyway, saints and sinners alike.

Since he accepts that God is only a postulate, Drees is free to reject the Omega Point god. Similarly he rejects the deist "Zero Point God" who creates the universe and its laws, and then leaves it alone. Drees wants a God of the present, and argues that such a God is consonant with all the cosmologies formulated by science. Admitting that this is a want and not a proof, that God is unnecessary to explain the data, Drees argues that God's presence "shines through" in our desire for perfection and justice in a world apparently devoid of either.

Drees classifies three types of responses to the apparent absence of perfection in the world. In the first, we simply deny the imperfection. This he relates to the New Age movement and modern American pop psychology, in which all the ills of society are assumed to be simply a crisis of perception. All we need to do for life to be in perfect harmony with the world is to want it that way. We are capable of anything, from becoming rich to walking on fire. No work is required. If we think we can do it, then we will. Much of modern American charismatic Christianity has adopted this idea, discarding all this depressing talk of hellfire and brimstone.

The proposals of Tipler and others for ultimate harmony, fully within the material world but far in the future, are also seen as this type of response. So are the more modest visions of progress that most people hold—the idea that things will someday be better. Realistically, we have no basis to suppose that these are any more than wishes and fantasies. Perhaps things will get better; perhaps they never will. The very fact that this is undetermined gives us reason to hope that they can get better, if we ourselves make the effort. But it is consciously directed effort, not simply a felt want, that will make this happen.

These considerations make Drees's second response to apparent imperfection of the universe more helpful than it first appears. This response has been articulated by Nobel physicist Steven Weinberg, and represents a widely held view among scientists: "The more the universe seems comprehensible, the more it becomes pointless."[53] There is simply no sign of a grand design wherever we look. But as humans we can make our own design and, by our modest efforts in art, science, and other noble human activities, lift human life above the level of farce and give it some of the "grace of tragedy."

Weinberg has attempted to clarify his position, explaining, "I did not mean that science teaches us that the universe is pointless, but rather that the universe itself suggests no point."[54] On God, he says, "The more we refine our understanding of God to make the concept plausible, the more it seems pointless."[55]

Drees forthrightly admits that he has no argument against this position, except that it might paralyze the person who holds it.

However, it has not paralyzed Weinberg, the late Richard Feynman, and many others who have expressed similar views. Far from being constraining, the absence of belief has always had a great liberating effect on the intellect. What motivation can anyone have for dreaming up new ideas and working hard to put them into action if those ideas and those actions are naught but the external commands of a supreme being pulling all the strings in heaven? What point would there be in trying to make the world a better place, if we have no power to do so?

Drees personally prefers a third response. The world is sufficiently ambiguous that, while perfection and justice appear absent, they cannot be ruled out. So the postulate of God, as a present transcendent reality representing the locus of the possibilities of perfection and justice, can still be consistently asserted.

So Drees finally proposes a model for God that he claims is consonant with science and the other cosmologies discussed. God is immanent, that is, present in the processes of nature. God is temporally transcendent as both creator and perfecter. God is spatially transcendent, perhaps embedded in higher dimensions of space so he can be equidistant from all points, like the center of a sphere. God is the locus of values and possibilities, the beginning and end of perfection and justice. Finally, God is the source of actuality. Even if the universe appeared out of nothing, a quantum fluctuation in a vacuum, that nothing is something, and God is the source of that something: "Rather than seeking an understanding of divine action in the world, the world itself is understood as God's action."[56]

But God remains a postulate. By postulating God, we choose to affirm our belief in the basic goodness of the world. And since we have no evidence that God exists, individuals play an active role in making that choice. We are not forced to believe by evidence or duress.

Thus, Willem Drees has made an important contribution to a sensible dialog between science and theology. Understanding the science better than most theists, he has been able to define a basis for belief, but one that still requires the usual complement of faith. In the end, Drees has affirmed that God is an additional postulate not required by the data.

The ancient introduction of immaterial spirits were honest human attempts to explain the unexplained. Science has gradually replaced these explanations with ones relying on matter alone and uncovered no sign of spirits or gods. Despite his abhorrence of the concept, Drees still presents us with a God of the gaps as an explanation for the remaining mysteries of the universe. Humanity may someday explain these mysteries without recourse to the transcendent, as it has the many mysteries of the past. Drees asks, "Can one worship a hypothesis?" I find it very difficult to bow before that altar.

In *Faith of a Physicist*, Polkinghorne was quite explicit in rejecting even the remotest chance that we live in a purely natural, purposeless universe. "The strategy of the materialist atheists," he says, "is usually to claim that science is all, and that beauty and the rest are merely human constructs arising from the hard-wiring in our brains. I cannot accept so grotesquely impoverished a view of reality."[57] But he still interprets the world from an anthropocentric, theocentric perspective. Nonbelieving scientists such as Richard Feynman and Carl Sagan looked at the material universe with wonder and found material reality to be anything but grotesquely impoverished.

NOTES

1. This is a pun on "The Promise Keepers," a group of highly conservative Christian men who were very active in the Unites States in the 1990s.

2. John Polkinghorne, "Chaos Theory and Divine Action," in *Religion & Science: History, Method, Dialogue*, ed. W. Mark Richardson and Wesley J. Wildman (New York and London: Routledge, 1996), pp. 243–52.

3. John Polkinghorne, *Belief in God in the Age of Science* (New Haven and London: Yale University Press, 1998), pp. 85–86.

4. Ibid., p. 82.

5. Ibid., p. 88.

6. Ibid., p. 85.

7. Ibid., p. 21.

8. John Polkinghorne, "The Metaphysics of Divine Action," in *Chaos and Complexity: Scientific Perspectives on Divine Action*, ed. R. J. Russell, N. Murphy, and A. Peacocke (Vatican City: Vatican Observatory, 1995), pp. 147–56; Polkinghorne, "Chaos Theory and Divine Action," in *Religion & Science: History, Method, Dialogue*, ed. W. Mark Richardson and Wesley J. Wildman (New York and London: Routledge, 1996), pp. 243–52; Arthur Peacocke, *Theology for a Scientific Age* (London: SCM Press, 1993); "God's Interaction with the World: The Implications of Deterministic 'Chaos' and Interdependent Complexity" in *Chaos and Complexity*, pp. 263–87.

9. James Gleick, *Chaos: Making a New Science* (New York: Penguin Books, 1987), pp. 11–18.

10. Roger Lewin, *Complexity: Life at the Edge of Chaos* (New York: Macmillan, 1992), p. 100.

11. Gleick, *Chaos*.

12. Thomas F. Tracy, "Particular Providence and the God of the Gaps," in *Chaos and Complexity*, pp. 289–324.

13. Ilya Prigogine, *From Being to Becoming* (San Francisco: W. H. Freeman, 1980); Ilya Prigogine and Isabella Stengers, *Order Out of Chaos* (New York: Bantam, 1984). Note that Prigogine's frequent coauthor is Isabella Stengers, who is no relation to me.

14. Prigogine and Stengers, *Order Out of Chaos*; Peter Coveney and Roger Highfield, *The Arrow of Time* (New York: Fawcett Columbine, 1991).

15. The slight time asymmetry that is seen in a few rare elementary particle processes is very small, one part in a thousand, and does not forbid time reversibility. It just means that these reactions do not proceed in both time directions at exactly the same rate. They do proceed at the same rate, however, when particles are replaced by antiparticles and the process is viewed in a mirror. For more details see my *Timeless Reality*.

16. For a critical review of Prigogine's claims of a new scientific paradigm, see Jean Bricmont, "Science of Chaos, or Chaos in Science?" *Physicalia Magazine* 17 (1995): 159–208, reprinted in "The Flight from Science and Reason," ed. P. R. Gross, N. Levitt, and M. W. Lewis, *Annals of the New York Academy of Sciences* 775 (1996): 131–75.

17. Peacocke, *Theology for a Scientific Age*, p. 118.

18. John Briggs and F. David Peat, *Turbulent Mirror: An Illustrated Guide to Chaos Theory and the Science of Wholeness* (New York: Harper & Row, 1989); Coveney and Highfield, *The Arrow of Time*.

19. Polkinghorne, *Belief in God in the Age of Science*, p. 88.

20. Ibid., pp. 97–98.

21. Ibid., p. 128.

22. Penrose, *The Emperor's New Mind: Concerning Computers, Minds, and the Laws of Physics* (Oxford: Oxford University Press, 1989), p. 112.

23. Polkinghorne, *Belief in God in the Age of Science*, pp. 128–30.

24. For more discussion of Penrose Platonism, see chapter 12 of my *The Unconscious Quantum: Metaphysics in Modern Physics and Cosmology* (Amherst, N.Y.: Prometheus Books, 1995).

25. John Wilson, "Examining Peacocke's Plumage," *Christianity Today* (March 12, 2001).

26. Nancy Murphy, "Divine Action in the Natural Order: Buridan's Ass and Schrödinger's Cat," in *Chaos and Complexity*, pp. 325–57.

27. Ibid.

28. Fritoj Capra, *The Tao of Physics* (Boulder: Shambhala, 1975); Gary Zukav, *The Dancing Wu Li Masters: An Overview of the New Physics* (New York: Morrow, 1979); Deepak Chopra, *Ageless Body, Timeless Mind: The Quantum Alternative to Growing Old* (New York: Random House, 1993); Amit Goswami, *The Self-aware Universe: How Consciousness Creates the Material World* (New York: G. P. Putnam's Sons, 1993).

29. Kenneth R. Miller, *Finding Darwin's God: A Scientist's Search for a Common Ground between God and Evolution* (New York: HarperCollins, 1999), p. 213.

30. Ian G. Barbour, *Religion and Science* (San Francisco: HarperCollins, 1997), p. 216.

31. John F. Haught, *God after Darwin* (Boulder, Colo.: Westview Press, 2000), p. 120.

32. Dean L. Overman, *A Case against Accident and Self-Organization* (New York, Oxford: Rowman & Littlefield, 1997), p. 1.

33. Ibid.

34. Rebecca Flietstra, "A Misguided Quest for Proof, " *Christianity Today* 4, no. 5 (1998): 34.

35. Stanley L. Miller, "The Origin of Life," in *This Is Life: Essays in Modern Biology*, ed. Willis H. Johnson and William C. Steere (New York: Holt, Reinhart, and Winston, 1962), pp. 316–41.

36. Flietstra, "A Misguided Quest for Proof."

37. As quoted in William Dembski, "Is Intelligent Design a Form of Natural Theology?" *Metanexus: The Online Forum on Religion and Science* [online], www.meta-list.org/archives/fulldetails.asp? archiveid=3130& Listtype=Magazine [May 11, 2000].

38. Ibid.

39. Templeton Foundation [online], www.templeton.org.

40. Ibid.

41. Ibid.

42. Ibid.

43. Ibid.

44. Phillip E. Johnson and Howard van Till, "God and Evolution: An Exchange," *First Things* 34 (1993): 32–41.

45. David A. Staff, "Christian Orthodox on Man: God's Very Special Creation" [online], http://www.amesefc.org/sermons/sr 080402.htm.

46. Willem B. Drees, *Beyond the Big Bang: Quantum Cosmologies and God* (La Salle, Ill.: Open Court, 1990).

47. Willem B. Drees, "Gaps for God?" in *Chaos and Complexity*, pp. 223–37.

48. Ibid.

49. Drees, *Beyond the Big Bang*. Drees did not have Tipler's book as a reference in writing his own book, but relied on earlier Tipler papers.

50. For a comparative study of Teilhard and Darwin, see H. James Birx, *Interpreting Evolution: Darwin & Teilhard de Chardin* (Amherst, N.Y.: Prometheus Books, 1991).

51. John D. Barrow and Frank J. Tipler, *The Anthropic Cosmological Principle* (Oxford: Oxford University Press, 1986).

52. For my review of Tipler, see Victor J. Stenger, review of *The Physics of Immortality*, by Frank. J. Tipler, *Free Inquiry* 15 (1995): 54–55.

53. Steven Weinberg, *The First Three Minutes* (New York: BasicBooks), p. 155.

54. Steven Weinberg, *Dreams of a Final Theory* (New York: Pantheon Books, 1992), p. 255.

55. Ibid., p. 256.

56. Drees, "Gaps for God?"

57. John Polkinghorne, *The Faith of a Physicist* (Princeton: Princeton University Press, 1994), p. 56.

TWELVE

THE GODLESS UNIVERSE

The dread and darkness of the mind cannot be dispelled by the sunbeams, the shining shafts of day, but only by an understanding of the outward form and inner workings of nature.

—Lucretius, ca. 60 B.C.E.

THE UNVERIFIED ASSERTION

As I have emphasized throughout this book, the assertions about God made by the three great monotheistic religions and the contentions about other worlds transcending matter made by mystics and paranormalists are legitimate subjects of scientific inquiry. If God or any other transcendent entity affects physical events and human affairs as significantly as most of their believers think they do, then these effects should be empirically detectable and confirmable by established scientific methods.

A God who created the universe would be expected to leave some unambiguous imprint of that creation behind. Otherwise, how would anyone know about it, including believers? A God who has hidden his creation from us is not the Judeo-Christian-Islamic God. If a supernatural creation happened, then we should see signs of it in our study of the content and structure of the universe. We do not.

A God who is prayed to is necessarily a God who answers prayers. Otherwise, why bother? If at least some prayers were answered, then we should, in principle, be able to see the result—not just in anecdotes but in careful, controlled experiments. We do not.

A god who does not intervene in the world in any significant way cannot be disproved, but that is also not the God of most Christians, Jews, and Muslims.

In the previous chapters I have offered examples of the types of observations that would have provided support for traditional beliefs if unambiguously confirmed by science:

- observations that cannot be explained except by a supernatural creation of the universe
- events induced by prayer or other supernatural intervention
- extrasensory perception (ESP) or the ability of the human mind to move matter in ways that cannot be explained by any known physical means
- examples in which faith healing or other forms of spiritual therapies cured the ill
- mystical or religious experiences during which verifiable information was obtained that could not possibly have been in the mystics' brain circuits all along
- quotations from sacred scriptures that contain facts that the people who wrote them could not have known
- evidence that the Sun and Earth are considerably younger than, say, a hundred million years, making evolution impossible

Not all of these observations would directly imply the existence of a supernatural realm. For example, ESP or alternative healing powers might be found to involve some previously undiscovered but entirely natural force. Likewise, if life did not evolve, it still may have resulted from purely natural processes.

Of course, I have argued that the evidence for evolution is overwhelming and that the Sun and Earth are certainly billions of years old. These are listed here because they provide good examples of the kinds of observations that might have offered scientific support for traditional beliefs, had they turned out different. If evolution had been refuted rather than confirmed, or if the solar system had been shown to be young rather than old, we might still have had plausible grounds for the notion that humanity is a special creation.

One can conceive of a situation in which one or more of the observations listed above was solidly confirmed and the supernatural explanation happened to be far more likely than a natural one. Like any scientific explanation, it would always remain provisional. Until that situation changed, however, a scientific reason would exist for supernatural beliefs. In any case, this is not the current situation. Rather, natural explanations remain more likely for all empirical observations and no rational basis exists for supernatural beliefs.

RATIONAL ASSUMPTIONS

I have no doubt that the science theists who have been involved in the search for nonmaterial phenomena and have promoted the necessity for nonmaterial explanations for material phenomena will disagree with my characterizations of their claims. Likewise, the parapsychologists who maintain that they have convincing evidence for psychic forces will dissent from my conclusions. Nevertheless, no matter how vehemently they may object to what I have written here, investigators from neither group can credibly deny the paucity of clear, unambiguous signals in their data. They will offer up excuses for their failure to detect a convincing signal, rather than admit the possibility that the signal they seek simply does not exist. They will argue why their nonmaterial assumptions are still better than material ones, rather than acknowledge that the continuing absence of confirmatory evidence renders this conclusion dubious.

Parapsychologists often suggest that ESP may not work well in the presence of nearby skeptics. This is called the "observer effect" and is assumed to be a property of the psychic force. The more rational assumption is that skeptics are less likely than believers to accept poorly controlled experiments and that the psychic force simply does not exist.

The decrease in the significance of positive ESP results with time is termed the "decline effect" and, like the observer effect, is also assumed to be a property of the phenomenon. The more rational assumption is that the experiments become better controlled so that the number of false positives declines with the passage of time and that the phenomenon simply does not occur at any time. Or, given the fact that none of the results had high statistical significance to begin with, the effect could have been simply a statistical artifact that went away with more data.

The absence of any decrease of ESP powers with distance, as one expects from energy conservation, is called the "distance effect" and is explained away by suggesting that perhaps the power is supernatural and need not conserve energy. The more rational explanation is that ESP does not exist at any distance.

Proponents of faith healing and alternative medicine argue that these therapies may not function in the cold, impersonal environment of a clinic or laboratory. The more rational argument is that they work only in the minds of believers.

Those who promote the efficacy of prayer protest that its effects are difficult to measure since praying cannot be easily controlled. The more rational conclusion is that prayers have no effect whatsoever.

Mystics state that their experience of oneness with God and the universe cannot be described in scientific terms. The more rational statement is that this experience is all in their heads.

Creationists assert that no one was around to see the origin of the universe, so we cannot say it was not a miracle. The more rational assertion is that no one was around to say it was a miracle.

As we have seen, none of the empirical results so far published on paranormal or nonmaterial phenomena reaches the level of significance normally required in physics and the other sciences that search for extraordinary phenomena. In only one case, the Princeton random number experiment, does the reported statistical significance exceed the conventional physics standard of 10,000-to-1 odds against an effect at least as great being produced as a chance artifact. However, the Princeton experiment is unconvincing on other grounds, and has never been replicated at the same effect size and level of statistical significance. This is the case even though reports of positive results go back decades and the cost of repeating the experiment would be negligible compared to what is expended in conventional science.

Those who still insist that the evidence for paranormal or supernatural events is significant would have us bypass conventional criteria commonly applied to other extraordinary claims. Instead, they ask us to employ the much weaker criteria for ordinary claims made in fields such as medicine, which have other more pressing priorities than leisurely probing the nature of the universe. In particular, the assertion that psychic phenomena have been confirmed based on the metanalysis of a host of individually insignificant experiments has been shown to be untrustworthy at best and, most likely, wrong. Furthermore, no extraordinary phenomenon has ever been uncovered initially by metanalysis.

THE MATERIAL SCENARIO

I have presented the case for a self-contained, fully material universe—one that operates on its own without outside intervention. Here is a short summary of its basic features:

- Space and time have no boundaries—no beginning or end.
- Time has no preferred direction at the fundamental level.

Time's arrow, as commonly experienced, is just a statistical definition that applies only to many-body systems such as those of common human experience. The second law of thermodynamics is simply a definition of the arrow of time in such systems.

- Our material universe arose from a quantum fluctuation in the void at an arbitrary space-time point, which led to an exponential expansion (inflation) and then the big bang. The entropy of the universe was the maximum for its volume at the moment of that fluctuation, so any information that may have been present was destroyed.
- More than one such universe likely happened in this way. At least, we have no basis for assuming that a universe happened only once.
- The global conservation laws of physics and the principles of relativity are simply symmetry properties of the void that naturally extend to the universe of matter. These laws and principles are likely to be the same in all universes.
- None of these laws was violated by the appearance of matter from the void.
- Additional, nonglobal laws of physics resulted from local spontaneous symmetry breaking in the early universe. These laws would probably differ from universe to universe since they are accidental.
- Structures such as galaxies, stars, planets, and living organisms evolved from the complex material systems that formed under spontaneous symmetry breaking. The formation of order did not violate the second law of thermodynamics since the maximum allowable entropy of the universe increases as the universe expands, allowing increasing room for order to form.

Not every aspect of this scenario can be "proven," and some may be modified or revoked by future developments. However, this does not prevent us from proceeding further. The material scenario is not unsupported conjecture. It is not pulled from thin air, but is based on extrapolating from our best current knowledge. For example, inflationary cosmology is now solidly supported by the data, and this strongly implies an original void. Furthermore, the highly successful standard model of elementary particles and forces relies heavily on the notion of global symmetry and local symmetry breaking, and has already been widely used in building models of the early universe.

As I have emphasized, the purpose of presenting this scenario, in detail and based upon the best rational interpretation of the evidence available to us now, is to demonstrate that we require no ingredients other than matter to describe currently observed reality. That reality shows no signs of immaterial forms of energy, transcendent forces, or the special injection of information from outside the universe. The material scenario thus serves to refute the widespread claim that modern physics and cosmology have uncovered the existence of a creator.

While science is no more exempt from criticism than any other human endeavor, and scientists are far from perfect, any suggestion that most scientists dogmatically refuse to consider paranormal or theistic claims is wholly unfounded and defamatory. Academic scientists have been involved in much of this kind of work. While they fall outside the scientific mainstream, these claims have been given the same hearing as extraordinary claims in any scientific field. Indeed, some have received a far greater hearing than almost any other extraordinary proposal. No scientific field except parapsychology has experienced over 150 years of exclusively negative results without being dismissed as a lost cause. Well-established, standardized criteria have been applied to all previous claims, and any new claim that meets these criteria will be given a fair hearing. The evidence will be allowed to speak for itself.

The problem here is with the evidence—or lack of it. If the evidence for the weak nuclear force were as insignificant as the evidence for psychic or other nonmaterial forces, then physicists would not include the weak nuclear force in their theories. If the data supporting the efficacy of antibiotics were as insignificant as the data claimed to support the efficacy of prayer, faith healing, and most of alternative medicine, then physicians would not prescribe antibiotics.

THROUGH A GLASS DARKLY

I have tried to show respect for the group of scientifically savvy theologians I have termed the "premise keepers." Perhaps this designation is a bit too flippant, but I think that it accurately characterizes their position. The premise keepers are valiantly trying to maintain the premises of traditional religious teachings, particularly Christianity, while bringing them into line with discoveries of science that they do not challenge. Most significantly, they accept the apparently dominant role of chance in the universe.

Unfortunately, the goal of accommodating science and religion seems impossible to achieve without sacrificing a number of essential traditional religious teachings, especially the central position of humanity in a purposeful cosmos. The premise keepers do not explicitly discard this teaching, but a careful reading of their proposals indicates to what great lengths they must go to maintain it. For example, many premise keepers do not hold to the form of guided evolution to which the Catholic church and many scientist-believers subscribe, in which God occasionally pokes his finger into the works to make things come out right.

In a sense, this group of proscience theists differ only by degree from those antiscience theists who reject evolution outright. Guided evolution is not the scientific theory of evolution, and most premise keepers understand this. Some have suggested that the evolutionary path that has led to humanity was indeed filled with chance events, as evolution theory indicates, but that divine purpose would still have been achieved along some other path along which humanity never evolved!

The displacement of humans from the focus of existence is surely unacceptable to the great majority of Christians, Muslims, and Jews (not to mention most professors of the social sciences and humanities). Where the ideas of the premise keepers have become known to mainstream Christians, they have not been exactly welcomed with open arms.

Nevertheless, the premise keepers, backed financially by the Templeton Foundation, have labored mightily to situate science as the basis of their theologies. But other theists who hold the preservation of ancient myths more sacred than established knowledge have begun to usurp the language and authority of science. At this writing, a powerful group of American theists endeavor to restore the supremacy of traditional theology and invalidate the findings of science by deconstructing the scientific method. Backed by a number of well-funded private organizations of their own, such as the Center for Science and Culture within the Discovery Institute, they use as their "wedge" a wildly unsupported contention that science is to blame for the immoral behavior they perceive all around them. From some of the more extreme utterances of this group, one would think that the world was a better place in the Dark Ages, when religion ruled and ancient materialist ideas such as atoms and evolution lay hidden away on dusty scrolls.

In this book I have attempted to fire a few rounds back into what has, in our society, become the unassailable territory of religion. No doubt many will be offended by my assault on their deeply held beliefs. My purpose here has not been to offend, but to defend. Anyone who values truth is morally obliged to test these beliefs with the same rigor they test any other assertion. And if religion continues to attack science, science has the duty to respond.

Scientists certainly have their work cut out for them if they are to convince the public that the "light of reason" is superior to the "light of faith." Many on the side of science advocate a cautious approach that does not overtly challenge religious teachings because of the fear of a public backlash. Some fear a disruption of the primary mechanism by which scientific endeavors are financed—public funding. Perhaps nonconfrontation would be the wisest short-term strategy, but I think scientists have been silent too long in allowing antiscience to proceed unchallenged. The devaluation of science can only hurt society in the long run.

People like what they see when they look in the mirror illuminated by the light of faith. It reflects an image of themselves as fallen angels, set on this planet with the divine purpose of rehabilitating themselves so they may rejoin their fellow angels in paradise. Unfortunately, the universe exposed by the light of science does not reveal a special place for humanity in the cosmos or any prospect for life after death. I would not be honest if I tried to sugarcoat these facts, just because they conflict so dramatically with common yearnings.

> When I was a child, I thought as a child, I understood as a child: but when I became a man, I put away childish things. For now we see through a glass darkly, but then face to face: now I know in part; but then shall I know even as also I am known (1 Cor. 13: 11–12, King James Version).

In an e-mail communication to my discussion group, avoid-L, Anne O'Reilly has reinterpreted these famous words of St. Paul so that they have greater relevance today:

> While faith illuminates the glass, it does so by turning it into a mirror; the glass is opaque and all we can see are the reflections of our own desires. The last three centuries of science have enabled us to penetrate this darkness and transform the mirror into a window on the universe. Now when we look out, the opaque glass has become translucent; with patches of transparency appearing. We begin to see what is truly out there; we come face-to-face with the terrible beauty of the cosmos that gave birth to us.

We can be sure that Paul did not intend his words to be so interpreted, but great poetry often carries with it truths that the poet did not comprehend when writing the original lines. Perhaps this is a sign that our deepest thoughts are not conscious ones.

Another avoid-L member, Ed Weinmann, has added:

> The feeling of life worthwhile, or of life worthless, is not dependent on buying into religious doctrines. It is a "feeling," pro-

duced by the interplay of many factors, all of which can be identified as material, and natural. Humanity is very good at procuring what it needs to survive, just like other animals. If humans need religion, they invent it. If other varieties of religion are needed, they, too, are invented.

Humanity has moved beyond childhood. We no longer need to depend on imaginary friends for company or a mythical sky-father to provide for our needs. We can take care of ourselves. We can find a way to live our lives that is consistent with the universe revealed to us by reason and science. While believers must struggle under the constant burden of guilt—because they can never live up to their cosmic status as fallen angels—nonbelievers can take pride in their achievements and those of the rest of our species. The great works of art, literature, and science are the result of our own efforts, not something inspired from another realm. Ideas such as democracy and liberty are human ideas, not God's. Our individual fates and the future of humankind are not already written, in either the laws of nature or the mind of God. These fates are in our hands. These hands, and the mighty potential that has evolved within our purely material brains, provide all the power we need to continue the upward advance of the human race.

The universe is not populated by mysterious forces, beyond our comprehension, that control our lives and destinies for some unseen purpose. Rather, thanks to science, humanity is in control and defines its own purpose.

APPENDICES

APPENDIX A: THE PLANCK LENGTH

By international agreement, the distance or length, L, between two points is defined as the time, t, it takes for light to travel between the points in a vacuum, multiplied by a constant, c:

$$L = ct \qquad\qquad (A1)$$

While c is called the speed of light in a vacuum, it is an arbitrary number that simply sets the units of distance. For example, if t is measured in seconds and you want L in meters, then $c = 3 \times 10^8$ by definition. Physicists and astronomers often work in units where $c = 1$, for example, when t is in years and L in light-years.

In order to measure t, we need a clock with an uncertainty, Δt, no larger than t. The time-energy uncertainty principle says that the product of Δt and the uncertainty in a measurement of energy in that time interval, ΔE, can be no less than $\hbar/2$, where $\hbar = h/2\pi$ and $h = 6.63 \times 10^{-34}$ Joule-sec is Planck's constant.

$$\Delta E\ \Delta t \geq \hbar/2 \tag{A2}$$

Thus,

$$\Delta E \geq \hbar/2t \geq \hbar c/2L \tag{A3}$$

This energy equals the rest energy of a body of mass m,

$$\Delta E = mc^2 \tag{A4}$$

Let L be the radius of a sphere. Within a spherical region of space of radius L we cannot determine, by any measurement, that it contains a mass less than

$$m = \hbar/2cL \tag{A5}$$

Now, let us consider the special case where the gravitational potential energy of a spherical body of mass m and radius R equals half its rest energy,

$$mc^2/2 = Gm^2/R \tag{A6}$$

so that,

$$R = 2Gm/c^2 \tag{A7}$$

This is called the *Schwarzschild radius*. Any body of mass m with radius less than R is a black hole. Suppose that $L = R$. Let us call that special case L_{PL}. Then, from (A5) and (A7),

$$L_{PL} = (\hbar G/c^3)^{1/2} \tag{A8}$$

which is called the *Planck length*, $L_{PL} = 1.6 \times 10^{-35}$ meter. We can see that it represents the smallest length that can be operationally defined, that is, defined in terms of measurements that can be made with clocks and other instruments. If we tried to measure a smaller distance, the time interval would be smaller, the uncertainty in rest energy larger, the uncertainty in mass larger, and the region of space would be experimentally indistinguishable from a black hole. Since nothing inside a black hole can climb outside its gravitational field, we cannot see inside and thus cannot make smaller measurement of distance.

Similarly, we can make no smaller measurement of time than the *Planck time*,

$$t_{PL} = L_{PL}/c = (\hbar G/c^5)^{1/2} \tag{A9}$$

which has the value $t_{PL} = 5.4 \times 10^{-44}$ second. Also of some interest are the *Planck mass*, which from (A5) is

$$m_{PL} = \hbar/cL_{PL} = (\hbar c/G)^{1/2} \tag{A10}$$

and has a value of 2.2×10^{-8} kilogram, and the Planck energy,

$$E_{PL} = m_{PL}c^2 = (\hbar c^3/4G)^{1/2} \tag{A11}$$

which has a value of 2.0×10^9 Joules or 1.2×10^{28} electron-volts. These represent the uncertainty in rest mass and rest energy within the space of a Planck sphere or within a time interval equal to the Planck time.

APPENDIX B: THE LIFETIMES OF STARS

The minimum lifetime of the class of stars larger than the Sun that can end their lives as supernovae can be calculated from

$$t_s = (\alpha^2/\alpha_G)\,(m_p/m_e)^2\,\hbar\,(m_p\,c^2)^{-1} \tag{B1}$$

where $\alpha = e^2/4\pi\varepsilon_0\hbar/c$ is the dimensionless strength of the electromagnetic force, e is the unit electric charge, ε_0 is the permittivity of free space (Standard International Units are being used), and

$$\alpha_G = G\,m_p^2\,(\hbar c)^{-1} \tag{B2}$$

is the dimensionless strength of the gravitational force, m_p is the mass of the proton, and m_e is the mass of the electron.[*] In our universe, $\alpha = 1/137$, $m_p = 1.67 \times 10^{-27}$ kilogram, $m_e = 9.11 \times 10^{-31}$ kilogram, $\alpha_G = 5.9 \times 10^{-39}$, and $t_s = 680$ million years. Actually, most stars, like our Sun, have much greater lifetimes; larger stars evolve more rapidly. The shorter-lived stars we are considering here generate the heavy elements from which planets and life later can evolve.

As we saw in appendix A, c is an arbitrary number that simply determines the units we use to measure physical quantities. The same can be said for $\hbar$ and G.[†] Indeed, each can be set equal to unity without changing any observational results. This does not mean that the strength of the gravitational force is arbitrary, since α_G depends on m_p, as seen in (B2). Thus, just three parameters determine the value of t_s: α, m_p, and m_e. We can choose any two to be whatever value we wish and then find the range of values for the third parameter that will give t_s equal to or greater than some value, say the value in our universe.

For example, suppose we are interested in a particular t_s and

---

[*]E. E. Salpeter, "Accretion of Interstellar Matter by Massive Objects," *Astrophysical Journal* 140 (1964): 796–800.

[†]For a further discussion of this point, see my *Timeless Reality: Symmetry, Simplicity, and Multiple Universes* (Amherst, N.Y.: Prometheus Books, 2000), chap. 15.

pick arbitrary values of m_p and m_e. Then, with the help of (B2), we can rewrite (B1)

$$\alpha = (Gm_p\, c\, t_s\,)^{1/2}\, m_e/\cancel{k}\qquad\qquad\text{(B3)}$$

to find the value of α needed. Note that no fine tuning is required to produce a universe with long-lifetime stars. We can always find a value of α for any value of t_s, equal to or larger than its value in our universe, regardless of the masses of the proton and electron. While the values of the three parameters may not yield some form of life for another reason, that reason is not insufficient time for stars to cook up the elements needed for life.

APPENDIX C: THE ENTROPY OF THE EXPANDING UNIVERSE

Entropy is defined by

$$S = k \ln\{\text{number of states}\} \qquad \text{(C1)}$$

where k is Boltzmann's constant. For N particles of the same type,

$$
\begin{aligned}
S &= k \ln\{(\text{number of states})^N\} \qquad \text{(C2)} \\
&= kN \ln \{\text{a not-too-big-number}\} \\
&\approx kN \\
&\approx N
\end{aligned}
$$

in units where $k = 1$.

Just after the initiation of the big bang, we can treat the universe as a sphere of radius R containing an expanding relativistic gas of total energy E. Its maximum entropy was equal to the maximum number of particles,

$$S_U{}^{\max} = E/\varepsilon_{\min} = E\,\lambda_{\max}/hc = 2\pi RE/hc = RE \qquad \text{(C3)}$$

where $\varepsilon_{\min} = hc/\lambda_{\max}$ is the minimum energy of the relativistic particles, and $\lambda_{\max} = 2\pi R$ is their corresponding maximum wavelength. I have assumed units where $\hbar = h/2\pi = c = 1$, which greatly simplifies the calculations. Note that, in these units, the Planck length given in (A8) is simply $L_{PL} = \sqrt{G}$. This equals the Planck time given in (A9), while the Planck mass and energy, from (A10) and (A11) are both equal to $1/2\sqrt{G}$.

The maximum entropy of a black hole of radius R and mass M, or rest energy Mc^2, is given similarly to (C3) by

$$S_{BH}{}^{\max} = Mc^2/\varepsilon_{\min} = E\,\lambda_{\max}/hc = 2\pi R\, Mc^2/hc = RM \qquad \text{(C4)}$$

From (A7), $M = R/2G$ and from (A8), $G = L_{PL}{}^2$. Thus,

$$S_{BH}{}^{max} = R^2/2L_{PL}{}^2 \qquad (C5)$$

Now let us consider the situation when the universe was a Planck sphere, $R = L_{PL}$. At that time, the universe is a black hole. Its maximum energy in that case would be the Planck energy (A11), $E_{PL} = 1/2\sqrt{G} = L_{PL}/2$ and so, from (C3),

$$S_U{}^{max} = R/2L_{PL} \qquad (C6)$$

which equals (C5) when $R = L_{PL}$. So, we see that, as the universe expands, $S_U{}^{max}$ increases linearly with R while $S_{BH}{}^{max}$ increases quadratically. Thus, although the universe starts out with maximum entropy, its maximum entropy becomes less than its maximum allowable entropy, that of a black hole of the same size, leaving increasing room for order to form. Since the universe has now, 13 billion years later, expanded to more than sixty orders of magnitude larger than the Planck length, its maximum entropy is at least sixty orders of magnitude below its allowable upper limit.

INDEX

Abraham, 224
absolute space, 205
acceleration, 202, 205
acceleration of the universe, 157
acupressure, 265
acupuncture, 263, 265, 268, 276
Adam and Eve, 167
Adams, John, 89
aether, 207, 275
aether waves, 266
age of Earth, 49
age of the Sun, 50
AIDS, 251
Albacete, Monsignor Lorenzo, 11
Alexander VI, 35
Allen, Woody, 222
alternative medicine, 21–22, 263, 266,
    268, 273, 275–76, 288, 342, 346
American Civil Liberties Union
    (ACLU), 10, 57, 59

American Society for Psychical
    Research, 278
amino acids, 154, 324
Anglican Church, 46
angular momentum, 193
animal magnetism, 265
Annals of Internal Medicine, 243
anthropic coincidences, 86–87, 153,
    156, 183
Apollo 13, 229
apologetics, 45, 102, 171
Aquinas, St. Thomas, 88
Archives of Family Medicine, 243
Archives of Internal Medicine, 245,
    249
argument from design, 67, 88, 93,
    99–100, 118, 149, 153, 159, 169,
    174, 219, 227, 236
Aristotle, 79, 88, 138, 159, 192, 264
Armstrong, Herbert, 33

Armstrong, Karen, 16
arrow of time, 41, 175–76, 179–80, 188, 344
artificial intelligence, 119
artificial life, 116
Asser, Seth, 241
astral projection, 291
astrology, 61
atomic hydrogen, 144
atomic theory of matter, 172, 267, 330
Augustine, St., 174, 190
aura, 270
Austin, James, 292
Avalos, Hector, 255
Ayala, Francisco, 57

backward causality, 177, 282
Bagiella, E., 243
Barbour, Ian, 321, 325–26
Barnum, P. T., 276
Baronius, Cardinal, 79
Barrow, John, 87, 332
Battle for God, The (Armstrong), 16
Baylor Institute of Faith and Learning, 101
Baylor Project on Religion and Science, 101
Baylor University, 100–101
Bayonne, New Jersey, 29
BBC News, 298
Beagle, voyage of, 49
Begley, Sharon, 13, 86, 293
Behe, Michael, 116, 117, 119, 236
believer effect, 283
Bell, John, 38
Bell Labs, 103
Benor, Daniel J., 254
Benson, Herbert, 246

Bentley, Richard, 88, 131, 138
Berlin Wall, 38
Bernoulli's principle, 61
Bible, 165, 167, 170, 224–25
Bible as a scientific document, 225
Bible Code, The (Drosnin), 227–28, 230–31, 232, 234, 236, 238, 255
Bible Code II, The (Drosnin), 236
biblical flood, 169
biblical prophecy, 221, 223, 229
Bienfield, Harriet, 252
bifurcation, 121
big bang, 13, 37, 56, 84, 86, 145, 158, 163, 167, 170–72, 174, 180, 189, 344
big crunch, 180
Bikini atoll, 35
bioenergetic field, 268–69, 275–76
bioenergy medicine, 269
blackbody radiation, 270–72
black hole, 144, 152, 189
Blackmore, Susan, 297, 299
blueshift, 168
Boeing 747, 113
Bohr, Niels, 274
Boltzmann, Ludwig, 176
Boltzmann's constant, 105, 356
Boltzmann's H-theorem, 105, 198
Bondi, Hermann, 171
Bonestall, Chesley, 36
Borel, Emile, 112
Borgia, Lucrezia, 35
Born, Max, 274
boron, 154
bosons, 143
Bowling Green State University, 245
brain death, 291, 298
brain waves, 272

Branch Davidians, 101
breath of the gods, 264
Bridgewater treatises, 325
Brief History of Time (Hawking), 86
broken symmetry, 196, 211, 216
Bronson, Po, 251–52
Brookhaven National Laboratory, 142, 145, 182
Bruyere, Rosalyn, 252
Bryan,William Jennings, 52
Buckland, William, 124
Buddha, 290
Burbidge, Geoffrey, 171
butterfly effect, 311–12
Byrd, Randolph, 247–49

calculus, 133–34, 139
Capra, Fritjof, 274
carbon, 153
cardiopulmonary resuscitation (CPR), 298
Carnap, Rudolf, 222
Case Against Accident and Self-Organization, The (Overman), 322
cathode rays, 279
Catholic Church, 35, 47, 346
CBS Early Show, 10
CBS News, 298
Center for Alternative Medicine, 267
Center for Science and Culture, 15, 64, 347
Center for the Study of Religion/ Spirituality and Health, 243
Center for Theology and the Natural Sciences, 13–14
centrifugal force, 205
cerebral anoxia, 292
cerebral hypoxia, 292
Chandra X-ray space telescope, 143

channelers, 277
Chanukah, 232
chaos, 181, 189
chaos theology, 310
chaos theory, 13, 311, 313–14, 316
charmed quarks, 38
chi, 264–65
Chinese medicine, 264
chiropractics, 268
Christianity Today, 315, 323
Cicero, 88
classical mechanics, 120, 199, 208
clinical death, 298
clockwork universe, 88, 90
cognitive theory, 22
Cold War, 38
Cole, Henry, 54
Comings, Mark, 252
Committee for the Scientific Investigation into Claims of the Paranormal (CSICOP), 288
commutative law, 209
Complementary Healing Research, 254
complementary medicine. See alternative medicine
complexity, 119, 121–22, 154, 184, 188, 196, 211, 213
complexity theory, 311, 314
complex specified information (CSI), 112, 238
consciousness, 314
Conscious Universe, The (Radin), 286
conservation of angular momentum 163, 198, 200, 209
conservation of energy, 39, 147–48, 163, 184, 192, 198, 200, 209, 312
conservation of information, 106, 109–10, 122

conservation of linear momentum, 163, 192, 198, 200, 209–10
conservation principles, 41, 344
Contact (movie), 113
continuous symmetry, 195
Copernican theory, 45, 47
Copernicus, Nicolaus, 24, 81, 125, 170, 197, 200
Coriolis force, 205
corona discharge, 273
cosmic microwave background (CMB), 84, 145–47, 151, 165, 222
Cosmic Microwave Background Explorer (COBE), 84, 146
cosmic rays, 120, 182
cosmological constant, 156, 158–59, 178, 182
cosmological constant problem, 156, 159
cost function, 114
Courcey, Kevin, 245
Cowan, Clyde, 222
Craig, William Lane, 172–74
creation/evolution war, 51, 64, 71
creationism, 51, 55, 60–61, 66–67, 172
Creation Research Society (CRS), 53
creation science, 53–54, 68, 94
Creator and the Cosmos, The (Ross), 83
Cro-Magnons, 167
Crookes, William, 266, 277, 279–80
Curie point, 216
curvature of space, 156–57, 182

Damascus, 224
D'Aquili, Eugene, 293

Dark Ages, 347
dark energy, 147, 157–58
dark matter, 147, 157
Darrow, Clarence, 52
Darwin, Charles, 15, 37, 44–45, 49–50, 56, 64, 67, 94, 124, 140, 328
Darwinism, 44, 47, 63–64, 94, 100, 122, 320, 328
Darwin on Trial (Johnson), 65
Darwin's Black Box (Behe), 116
Davies, Paul, 91, 189
da Vinci, Leonardo, 133
Dawkins, Richard, 43, 80, 94, 113, 142
de Broglie, Louis, 274
de Broglie wavelength, 120
Declaration of Independence, 90
decline effect, 342
deism, 23, 37, 88–92, 308, 317
Dembski, William, 63–64, 99–104, 106, 108–10, 112, 114–17, 119, 121, 233, 236, 238, 322–27
de Medici, Cosimo, 135
Democritus, 41, 191
Dennett, Daniel, 45
Descartes, René, 37, 263
design theory, 21
de Sitter solution, 182
deterministic chaos, 316
Deuteronomy, 233–34
Diana, Princess, 232
Dionysus, 190
Dirac, Paul, 274
Discovery Institute, 15, 18, 64, 387
discrete symmetry, 195
distance effect, 342
distant healing, 251
DNA, 49, 84, 113, 318, 329

doppler effect, 144

Dossey, Larry, 254–55, 297

double blind procedure, 251, 253

Drees, Willem B., 163, 331–34

Drosnin, Michael, 228–30, 232–36, 238

Duke University, 243, 279–80

Dunne, Brenda, 282

Easterbook, Gregg, 14

Eastern mysticism, 319

E. Coli bacteria, 113

Eddington, Sir Arthur, 176, 221

Edis, Taner, 122

Edwards v. Aguillard, 58, 232

Egerton, Rev. Francis Henry, 325

Egypt, 224

Einstein, Albert, 23, 37, 85, 87, 91–92, 156, 177–78, 181–83, 191, 203–205, 207, 221, 229, 274–75, 309

élan vital, 263

Eldredge, Niles, 312

electric charge, 164, 193

electroencephalography, 291–92

electromagnetism, 194, 203, 216, 275, 279

elementary particles, 38, 143

Ellis, George, 13

emergent phenomena, 310

Emperor's New Mind, The (Penrose), 119

energy of the universe, 148, 180, 188

Enlightenment, age of the, 23, 37, 88–89, 92, 308, 317

entropy, 104–106, 108–109, 149, 152, 164, 176, 180–81, 189, 213–14, 313

entropy of a black hole, 356

entropy of the universe, 344, 356

epidemiology of religion, 241–42

epilepsy, 292

equidistant letter sequences (ELS), 228

Erasmus, Desiderius, 219

ESP cards, 279, 286

Euclidean space, 207–208

Euclidean universe, 158

European Organization for Nuclear Research (CERN), 143

evolution, 48, 50–51, 55, 63–64, 66, 69, 83, 94, 100, 121, 312, 341, 346–47

evolutionary algorithms, 115

evolution theology, 320, 324

extrasensory perception (ESP), 39, 273, 279, 287, 340, 342

Ezekiel, 224

"fail safe file drawer" technique, 289

faith healing, 240, 342, 346

Faith of a Physicist, The (Polkinghorne), 335

Fallwell, Jerry, 10

false vacuum, 217

falsifiability, 60, 222

Feng shui, 265

Fermi, Enrico, 222

ferromagnetism, 214, 216

Feynman, Richard, 38, 178, 334–35

fictitious forces, 205, 207

fields, 194, 207

filedrawer effect, 232, 235, 250, 285–90

"Final Anthropic Principle" (Tipler), 332

Finding Darwin's God (Miller), 320
fine-tuning, argument from, 86, 153, 159, 183, 227, 236, 309, 331
Fingerprint of God, The (Ross), 83
firmament, 169
First Cause, 88
First Evangelical Free Church in Ames, Iowa, 330
first law of thermodynamics, 39, 147, 163
First Things, 328
fitness function, 114
fitness landscape, 114
Flietstra, Rebecca, 323–24
Founding Fathers, 90
four-dimensional space-time, 204
Fox, Kate, 276–77
frame of reference, 200, 208
Franklin, Benjamin, 89–90
Freeman, Chris, 298
Frontline, 11

Galilean relativity, 202
Galilei, Galileo, 45, 47, 79, 125, 131, 138–40, 194, 200, 202–204
Galle, Johann Gottfried, 221
Galton, Francis, 242
Gans, Harold, 230
Ganzfeld experiments, 289
Gates, Bill, 232
general theory of relativity, 91, 144, 156, 172, 178, 181–82, 205, 207–209, 221
Genesis, 46, 53–54, 165–68, 224–25, 227–28, 230, 232, 330
Genesis Flood, The (Morris and Whitcomb), 53
Genesis Question, The (Ross), 83
genome, 114

geodesic, 208
Gibbs, Josiah Willard, 105
Gish, Duane T., 55–56, 60, 62, 70–71
Glanz, James, 141–42
glioblastoma multiforme (GBM), 251
global symmetry, 184, 345
gluons, 38, 143
Glynn, Patrick, 19
God After Darwin (Haught), 321
God of the gaps, 122–24, 132, 159, 331, 335
God: The Evidence (Glynn), 19
Goethe, Johann Wolfgang, 307
Gold, Thomas, 171
Goldstein, Rabbi Herbert, 91
Good, I. J., 287
Gordon, Bruce, 101
Gospel of John, 67
Gould, Stephen Jay, 57, 79–80, 312
Graham, Billy, 10
gravitational mass, 275
gravitational pressure, 183
gravitational repulsion, 156
gravity, 155, 181, 194, 204–205, 207, 354
Greyson scale, 298
Griesediek, Rev. Joseph, 11
Grünbaum, Adolf, 86

H-theorem, 105
Halley, Edmund, 136, 221
Halley's comet, 136, 223
Hamilton-Jacobi equation of motion, 209
Harary, Keith, 251
Haught, John, 321, 328
Hawking, Stephen, 86, 91, 147, 191
Hayden Planetarium, 36
Hazor, 224

Healing Words (Dossey), 253
heat death of the universe, 149
heavy elements, 153
Hebrew Bible, 233
Hebrew calendar, 234
Hebrew University, 227, 231
Heisenberg, Werner, 90, 181, 189, 209, 274, 317
helium, 153
hidden variables, 177
Hinduism, 91
Hiroshima, 229
Hirschfield, Rabbi Brad, 11
Hitler, Adolf, 232, 261
holism, 268–69, 274–75, 314
holistic healing, 274
Holocaust, 229, 234
Home, Daniel Dunglas, 277
homeopathy, 267
Honorton, Charles, 289
Houck, Jack, 287
Houdini, Harry, 277
Hoyle, Fred, 84, 171
Hubble, Edwin, 145, 156
Hubble expansion, 168
human energy field, 21, 263, 270, 309
Humani Generis (Pius XII), 47
Hume, David, 93–95, 100
Humphrey, Nicholas, 249–50
Huxley, Thomas, 46
hydrogen, 153

inertial mass, 275
inertial reference frames, 205
inflationary cosmology, 145, 148, 157–58, 178, 180–82, 217, 222, 344–45
information, 103–104, 106, 108–10, 112–13, 122, 211, 312, 344

information theory, 20, 102–103, 106
instability, 311
instability of the void, 212
Institute for Creation Research (ICR), 53, 56
intelligent design (ID), 16, 62, 94, 99–100, 102, 106, 110, 113, 118, 122, 124, 159, 236, 324, 326, 328
intercessory prayer, 246–47
Internet, 19, 131
intersubjectivity, 212
invariance, 208
irreducible complexity, 116
Isaiah, 224, 228
Islam, 9, 172
Israel, 229, 234–35

Jackson, Allyn, 233, 235
Jahn, Robert G., 281, 282
Jefferson, Thomas, 23, 89
Jefferys, William, 282
Jeremiah, 224
Jerusalem Bible, 234
Jesus, 224, 232, 236–37, 290
John Paul II, 48
Johnson, George, 13
Johnson, Phillip, 64, 66–69, 102, 328–30
Jordan, Pascual, 274

Kabbalistic tradition, 227
Kaläm cosmological argument, 172–73, 179, 183
Kass, Robert E., 228, 231
Kelvin, Lord. *See* Thomson, William Lord Kelvin
Kelvin absolute temperature, 50
Kennedy, John F., 229, 232, 239
Kennedy, Robert F., 229, 239

Kentucky Derby, 277
Kepler, Johannes, 139
ki, 264
King James Bible, 224
King Jr., Martin Luther, 232, 239
Kirlian, Semyon Davidovich, 272
Kirlian aura, 263, 272–73
Koenig, Harold G., 243–45
Koons, Rob, 108
Koran, 165
Koren version of Hebrew Bible, 227, 233
Koresh, David, 229
Kuhn, Thomas, 43

Laplace, Pierre Simon, 123, 317
Large Hadron Collider (LHC), 143
Large Magellanic Cloud, 222
Laudan, Larry, 60–62
laws of mechanics, 123, 265
laws of motion, 200, 205
Leaning Tower of Pisa, 132
Leibniz, Gottfried Wilhelm, 131, 138–39, 205
Lemaître, Georges, 84
Lemon test, 57, 59
Leningrad codex, 233
Lerner, Eric, 171–72
Leslie, John, 309
Leucippus, 191
Levashov, Nicolai, 252
Leverrier, Urbain Jean Joseph, 221
life, definiton of, 154
Life After Life (Moody), 290
life force, 275
lifetime of stars, 155
Limits of Science, The (Medawar), 106
Lincoln, Abraham, 232
Linde, Andre, 183

lithium, 153
local symmetry breaking, 345
Lodge, Oliver, 266, 277, 279–80
logistic map, 312
logos, 16, 19, 22, 24, 290
Lorentz, Hendrick, 204, 216
Lorentz transformation, 203–204
Lorenz, Edward, 311
Lotz, Anne Graham, 10
Louisiana Tangipahoa Board of Education, 59
Lucretius, 41, 137, 187, 339
Luke, St., 224
Lyell, Charles, 49, 124

M-theory, 125, 191, 193
Mach, Ernst, 205
Mach's principle, 205
Maddox, John, 171
Madison, James, 89
Mafia, 35
magnetic dipole moment, 271
Mandelbrot set, 315–16
many worlds interpretation of quantum mechanics, 177, 188
Mark, St., 224
materialism, 66
mathematical theory of communication, 103
Matthew, St., 224
Matthews, Dale, 243
Maxwell, James Clerk, 216
McCleod, Owen, 79
McKay, Brendan, 230–32, 235
McVeigh, Timothy, 229
Medawar, Peter, 106
Mencken, H. L., 53
metanalyses, 285–89, 343
meter, definition of, 193

metric tensor, 191
Meyer, Stephen, 325
Michelson, Albert, 275
Microwave Anisotropy Probe (MAP, now WMAP), 146
Middle East, 234
migraine, 292
Milky Way, 145, 222
Miller, Keith, 70
Miller, Kenneth, 65, 67, 117, 320, 328, 331
Miller, Stanley, 324
mind of God, 190
mind-over-matter, 263
Mind Race, The (Russell and Harary), 251
Minkowski, Hermann, 204
Mithraism, 190
Miwok people, 252
Moby Dick (Melville), 232
Mohammed, 290
Monod, Jacques, 77
Monte Carlo analysis, 288
Moody, Raymond, 290
Moon landing, 229
Moore, Dan II, 251–52
Morley, Edward, 275
Morris, Henry M., 53, 56, 58, 62, 170
Muller, H. J., 117
multiple universes (multiverse), 183–84, 188
multivariate analysis, 248
Murphy, Nancy, 317–18
mystical experiences, 290–96
mysticism, 293
mythos, 16, 22, 24, 67, 290

Narlikar, Jayant, 171
National Academy of Sciences, 78

National Institutes for Health, 267
naturalism, 64–69
natural law, 60
natural religion, 88
natural selection, 49, 63, 94, 113, 116
natural theology, 37, 93, 324–26
Nature magazine, 40, 171, 287
Nature of Mind, The (Stokes), 287
Nature of the Gods, The (Cicero), 88
Nazi, 232
Neanderthals, 167
near-death experience (NDE), 290, 292, 296–98
Nebuchadnezzar, 224
nebulae, 145
neo-Darwinian biology, 321
Neptune, 221
neural networks, 119
neuroscience, 22
neutrino, 222
neutrino astronomy, 38
neutrino mass, 38
neutron, 143
neutron star, 153
New Age, 263, 291, 319, 333
Newark College of Engineering, 37
Newberg, Andrew, 293–95
New Jersey Institute of Technology, 37
Newsweek, 13, 24, 86, 293
Newton, Sir Isaac, 37, 45, 88, 108, 123, 131, 138–39, 202, 205, 210, 265, 309, 317
Newtonian physics, 44, 88, 91, 123, 125, 202, 221, 265, 274
Newtonian world machine, 123, 317, 321
Newton's law of gravity, 136, 145, 265

Newton's laws of motion, 132–33, 145, 202–203
nihilism, 33
Nile River, 224
Nixon, Richard, 37
Noah's ark, 53
Noether, Emmy, 199
Noether's theorem, 199, 209
No Free Lunch (Dembski), 114–15
no free lunch theorems, 115
noncommuting operators, 209
non-Euclidean geometry, 207
noninertial reference frames, 205
nonlinearity, 311
nonoverlapping magisteria (NOMA), 79
Nostradamus, 222
Not By Design (Stenger), 39, 174
Notices of the American Mathematical Society, 232
nuclear energy, 84
nuclear fusion, 153

Oakes, Edward T., 67
observer effect, 283, 342
Occam's razor, 157, 179, 184
O'Connor, Cardinal, 91
Oklahoma bombing, 229
Omega Point god, 332
optimization problems, 114
O'Reilly, Anne, 348
orientation association area (OAA), 294
Origin of the Species (Darwin), 49
origin of the universe, 342
Orr, H. Alan, 115
oscillating universe, 179
Osiris, 190
Ostrander, Sheila, 273

Oswald, Lee Harvey, 223, 229
out-of-body experience (OBE), 291, 296–99
Overman, Dean L., 322–24
Overton, Judge William R., 57–61, 69

p-value, 226, 238, 244, 249, 251
Paine, Thomas, 89–90
Paley, William, 67, 93–94
Paley's watch, 94, 113, 116
palm reading, 61
pantheism, 23, 88, 91–92, 190, 309
paradigm shift, 43
parallel universes, 188
parapsychology, 22, 263, 266, 280, 283, 285, 287, 296, 345
Pargament, Kenneth, 245
Parnia, Sam, 298
particle physics, 193
Patterson, Elissa, 270, 272
Paul, St., 290
Pauli, Wolfgang, 222, 274
Peacocke, Arthur, 311–15, 326, 332
Pennock, Robert T., 65–67, 69
Penrose, Roger, 87, 119, 121, 191, 315
Penzias, Arno, 145
People for the American Way, 10
periodic table, 153
Persinger, Michael, 292
phase transitions, 213
photon, 271, 274
photon theory of light, 275
Physical Review, 40
Physics and Psychics (Stenger), 39, 254
Physics of Immortality, The (Tipler), 332

Pius IX, 47
Pius XII, 37, 47, 84, 164
Planck, Max, 271, 274
Planck energy, 353, 356
Planck length, 152, 181, 189, 351, 353, 356–57
Planck mass, 353, 356
Planck's constant, 352
Planck sphere, 181, 189
Planck time, 180–81, 189, 353, 356
plasma universe, 171
Platonism, 119, 174, 187, 189–91, 197, 315, 319
Pluto, 213
pneuma, 264
Poincaré symmetry, 204
Polanyi, Michael, 100–101
Polkinghorne, John, 308–15, 318, 326, 331, 335
Pontifical Academy of Sciences, 48, 84
Popper, Karl, 222
Posner, Gary, 248
Powell, T., 243
prana, 264
prayer studies, 21, 237, 241, 342, 346
Prigogine, Ilya, 313–14
Princeton Engineering Anomalies Research group (PEAR), 281–86, 343
Princeton University, 281
principle of equivalence, 205–206
principle of relativity, 200–208, 344
Principles of Geology (Lyell), 49, 124
properties of the void, 212
prophecy, 221
proton, 143, 155
pseudorandom numbers, 285
psi of psychic phenomena, 21, 254, 262, 274, 285–88, 299, 343

psyche, 264
psychedelic drugs, 291
psychic energy, force, 263, 266, 276, 283, 341–42, 345
psychic healing, 251
psychics, 277
psychic waves, 275
psychokinesis, 273
Public Broadcasting System, 11
punctuated equilibrium, 170, 312

qi, 264
quanta, 271
quantum chromodynamics (QCD), 143
quantum electrodynamics, 216
quantum fields, 191, 309
quantum field theory, 211
quantum fluctuation, 182–83, 189, 334, 344
quantum gravity, 180–81
quantum healing, 274
quantum interference, 120
quantum mechanics, 13, 40, 44, 84, 90, 148, 173, 180–81, 189, 199, 265, 269, 271, 273, 275
quantum mysticism, 273, 285, 319–20
quantum paradoxes, 41
quantum theology, 316, 320
quark, 143, 190
quark confinement, 168
quark-gluon plasma, 142
Quinn, Philip, 62
quintessence, 156, 159

Rabin, Yitzhak, 229–30, 232, 234, 238–39
Radin, Dean, 286, 289
Randi, James, 241
random mutation, 122

randomness, 119, 320
Raymo, Chet, 63
Reasons to Believe, 82
redshift, 144, 168
reductionism, 268, 275, 314
Rehnquist, Justice William, 59
Reines, Frederick, 222
relativity, 84, 184, 200
religious experiences, 290, 340
religious harm, 241, 246
rest mass, 275
Revelation, 232
Revised Standard Version (Bible), 235
Rhine, Joseph Banks, 279–80
Ring, Kenneth, 291, 297
Rips, Eliyahu, 227–28, 230–31, 233, 235
Robertson, Pat, 9
Roche, David, 121–22
Rocky Mountains, 114
Rodenhouse, Jason, 116
Rogers, Martha, 269
Rosenberg, Yoav, 227, 231
Ross, Hugh, 83–87, 92, 110, 124, 146, 154, 156, 159, 164, 170, 172, 236, 309
Roswell, New Mexico, 232
rotational symmetry, 204, 211, 216
Royal Edinburgh Hospital, 298
Royal Society of London, 325
ruach, 264
Ruby, Jack, 229
Ruse, Michael, 57–60, 62

Sadat, Anwar, 229, 239
Sagan, Carl, 113, 141, 335
Samhyka, 165
Sandage, Alan, 13
Scalia, Justice Anthonin, 59

Scargle, Jeffrey, 289
schizophrenia 292
Schmidt, Helmut, 284
Schrödinger, Erwin, 209, 274
Schrödinger equation, 209, 319
Schroeder, Gerald L., 165–70, 273
Schwarzschild radius, 352
"Science of Unitary Human Beings" (Rogers), 269
scientism, 25, 288
Scopes, John Thomas, 52
Scopes monkey trial, 52
Scott, Eugenie, 54
Scott, Phillip, 252
séance, 276–77
second law of thermodynamics, 39, 108–109, 122, 147, 149, 152, 164, 176, 213–14, 313, 344
September 11, 2001, 9
700 Club, 9
Shadows of the Mind (Penrose), 119
Shallit, Jeffrey, 116
Shannon, Claude, 103–105, 122, 150
Shannon entropy, 103
Shannon uncertainty, 103–106, 108–109
sharpshooter's fallacy, 253
Shermer, Michael, 77
Sicher, Fred, 251, 253
silicon, 154
simplicity, 188
simultaneity, 203
single photon emission computed tomography (SPECT), 294
Sistine Chapel, 35
Sixtus IV, 35
Skeptical Inquirer, 287–88
Skeptics Society, 77
Sloan, Richard, 101, 243–44

Sloan, Robert, 100
Smith, Helene, 174, 251
Smith, Quenton, 173
Smoot, George, 84
Society for Psychical Research, 278
soul, 263
Southampton University, 298
space rotation symmetry, 195, 199
space-time rotation symmetry, 204
space translation symmetry, 194, 197, 210
special relativity, 177, 203–204, 209
spectral lines, 144
speed of light, 177, 203
spin, 193
Spinoza, Baruch de, 23, 85, 91, 309
Spinoza's god, 91
spiritual healing, 270
spiritualism, 276
spontaneous symmetry breaking, 184, 192, 209, 211, 216, 344
Staff, David A., 330
standard model of elementary particles and forces, 41, 56, 143, 145, 191, 211, 216, 222, 345
statistical mechanics, 104–105
Statistical Science, 227, 231
statistical significance, 225, 283, 342
steady-state universe, 171
Stefanatos, Joan, 269, 274
Sternberg, Shlomo, 233
St. Mary's Catholic Church, Bayonne, New Jersey, 29
Stokes, Douglas, 287–89
string theory, 191. *See also* M-theory
strong nuclear force, 211
Structure of Scientific Revolutions (Kuhns), 43
Sulloway, Frank, 78

superconductivity, 120
supernovae, 153, 155, 203, 222
surprisal, 104
Sutera, Raymond, 71
Swan, Rita, 241
symmetries of the void, 200
symmetry principles, 197–98
synaptic gap, 120

Talmud, 233
Targ, Elisabeth, 250–52
Targ, Russell, 251
Tart, Charles, 296
Teilhard de Chardin, Pierre, 36, 332
Tel Aviv, Israel, 229
teleology, 36
Templeton, John, 14
Templeton Foundation, 14, 326, 347
Templeton Prize, 14, 91, 189, 308, 311, 315, 321
Tessman, Irwin, 248
Tessman, Jack, 248
theism, 88, 90–92
theistic naturalism, 329
theory of everything (TOE), 37, 191
theory of evolution, 124
theory of relativity, 40–41, 275
therapeutic touch, 263, 269, 276
thermal equilibrium, 149
thermodynamics, 50
Thomas, Dave, 232, 236
Thomas, Justice Clarence, 59
Thomson, William Lord Kelvin, 49–50
Tibetan meditation, 293
Timeless Reality (Stenger), 41, 153, 192, 210, 282, 313, 320
time reflection symmetry, 175, 179
time reversal, 175, 282

time translation symmetry, 196, 199

time travel, 282

Tipler, Frank, 87, 332–33

Tolman, Richard, 105

Tolstoy, Leo, 228

Tooth Fairy, 320

top-down causality, 313

Torah, 227, 230

torque, 210–11

touch therapy, 268

Tracy, Thomas, 313

transitional forms, 71

Trinity, doctrine of, 88

true vacuum, 217

Tryon, Edward, 182–83

uncertainty principle, 90, 181–82, 189, 318

Unconscious Quantum, The (Stenger), 40, 153, 320

Unified Biofield Theory, 268

U.S. National Institutes of Health's Center for Complementary and Alternative Medicine, 251

U.S. Supreme Court, 54, 58, 232

University of California at Los Angeles (UCLA), 37

Uranus, 221

Ussher, Bishop James, 167

Utts, Jessica, 280–81

Van Till, Howard, 113, 328

Vatican, 172

Vaughn, Alan, 287

vesicle, 120

vital energy, 263

vital force, 271

vitalism, 276

void, definition of, 192

void, properties of, 192, 209

void-field, 194

W-boson, 222

Waco, Texas, 100

Walker, Scott, 246

Wallace, Alfred Russel, 44, 49–50, 56, 94

War and Peace (Tolstoy), 228, 232

Washington, George, 89

Watergate, 229

wave function, 274, 319

wave function collapse, 319–20

weak nuclear force, 211, 345

Weaver, Charles, 100

Weaver, Warren, 103

Wedge of Truth, The (Johnson), 64

wedge strategy, 15, 64, 347

Wein, Richard, 115

Weinberg, Steven, 38, 44, 191, 333

Weinmann, Ed, 348

Weissmandel, Rabbi H. M. D., 228

Western Journal of Medicine, 251–52

Westwood Community Methodist Church, Los Angeles, 37

whales, evolution of, 70

Wheeler, Richard, 264

Whitcomb Jr., John C., 53

White, Andrew Dickson, 45

white dwarf, 153

Wilber, Ken, 274

Wilberforce, Bishop Samuel, 46–47

Wilson, John, 315

Wilson, Robert, 145

Wired magazine, 14, 251

Witztum, Doron, 227–33

World Trade Center, 11, 229

Worldwide Church of God, 33

World Wide Web, 25, 68

YHWH, 190
young-Earth creationists, 124

Z-boson, 222
Zechariah, 224

zen, 292
Zero Point god, 332
Zuckerman, Desda, 252

CPSIA information can be obtained
at www.ICGtesting.com
Printed in the USA
BVHW082312170721
612102BV00005B/7/J